I0884779

Selected Directed Energy Research and Development for U.S. Air Force Aircraft Applications

A Workshop Summary

Robert J. Katt, Rapporteur

Air Force Studies Board

Division on Engineering and Physical Sciences

NATIONAL RESEARCH COUNCIL
OF THE NATIONAL ACADEMIES

THE NATIONAL ACADEMIES PRESS
Washington, D.C.
www.nap.edu

THE NATIONAL ACADEMIES PRESS 500 Fifth Street, NW Washington, DC 20001

NOTICE: The project that is the subject of this report was approved by the Governing Board of the National Research Council, whose members are drawn from the councils of the National Academy of Sciences, the National Academy of Engineering, and the Institute of Medicine.

This is a report of work supported by Grant FA9550-12-1-0413 between the U.S. Air Force and the National Academy of Sciences. Any views expressed in this publication are those of the author(s) and do not necessarily reflect the views of the organizations or agencies that provided support for the project.

ISBN-13: 978-0-309-29261-0
ISBN-10: 0-309-29261-1

Copies of this report are available from:

The National Academies Press
500 Fifth Street, NW
Keck 360
Washington, DC 20001
(800) 624-6242 or (202) 334-3313
http://www.nap.edu

Printed in the United States of America

THE NATIONAL ACADEMIES

Advisers to the Nation on Science, Engineering, and Medicine

The **National Academy of Sciences** is a private, nonprofit, self-perpetuating society of distinguished scholars engaged in scientific and engineering research, dedicated to the furtherance of science and technology and to their use for the general welfare. Upon the authority of the charter granted to it by the Congress in 1863, the Academy has a mandate that requires it to advise the federal government on scientific and technical matters. Dr. Ralph J. Cicerone is president of the National Academy of Sciences.

The **National Academy of Engineering** was established in 1964, under the charter of the National Academy of Sciences, as a parallel organization of outstanding engineers. It is autonomous in its administration and in the selection of its members, sharing with the National Academy of Sciences the responsibility for advising the federal government. The National Academy of Engineering also sponsors engineering programs aimed at meeting national needs, encourages education and research, and recognizes the superior achievements of engineers. Dr. C. D. Mote, Jr. is president of the National Academy of Engineering.

The **Institute of Medicine** was established in 1970 by the National Academy of Sciences to secure the services of eminent members of appropriate professions in the examination of policy matters pertaining to the health of the public. The Institute acts under the responsibility given to the National Academy of Sciences by its congressional charter to be an adviser to the federal government and, upon its own initiative, to identify issues of medical care, research, and education. Dr. Harvey V. Fineberg is president of the Institute of Medicine.

The **National Research Council** was organized by the National Academy of Sciences in 1916 to associate the broad community of science and technology with the Academy's purposes of furthering knowledge and advising the federal government. Functioning in accordance with general policies determined by the Academy, the Council has become the principal operating agency of both the National Academy of Sciences and the National Academy of Engineering in providing services to the government, the public, and the scientific and engineering communities. The Council is administered jointly by both Academies and the Institute of Medicine. Dr. Ralph J. Cicerone and Dr. C. D. Mote, Jr. are chair and vice chair, respectively, of the National Research Council.

www.national-academies.org

COMMITTEE ON ASSESSMENT OF DIRECTED ENERGY RESEARCH AND DEVELOPMENT FOR U.S. AIR FORCE APPLICATIONS: A WORKSHOP

RONALD E. KEYS, RK Solution Enterprises, LLC, *Chair*
WILLIAM L. BAKER, Independent Consultant, Sandia Park, New Mexico
JACQUELINE G. GISH, Independent Consultant, Glendale, California
MATT L. MLEZIVA, Wildwood Strategic Concepts, LLC
THOMAS E. ROMESSER, Independent Consultant, Torrance, California
DAVID J. SCOTT, Independent Consultant, Payton, Colorado
FRANK J. SERNA, The Charles Stark Draper Laboratory, Inc.
ANDEW M. SESSLER, E.O Lawrence Berkeley National Laboratory

Staff

CARTER W. FORD, Program Officer
MARGUERITE E. SCHNEIDER, Administrative Coordinator

AIR FORCE STUDIES BOARD

Acknowledgment of Reviewers

This report has been reviewed in draft form by individuals chosen for their diverse perspectives and technical expertise, in accordance with procedures approved by the National Research Council's (NRC's) Report Review Committee. The purpose of this independent review is to provide candid and critical comments that will assist the institution in making its published report as sound as possible and to ensure that the report meets institutional standards for objectivity, evidence, and responsiveness to the study charge. The review comments and draft manuscript remain confidential to protect the integrity of the deliberative process. We wish to thank the following individuals for their review of this report:

William L. Baker, Independent Consultant, Sandia Park, New Mexico,
Ken Bruner, U.S. Pacific Command,
John H. Hammond, Independent Consultant, Fairfax, Virginia,
Lester L. Lyles, The Lyles Group,
Matt L. Mleziva, Wildwood Strategic Concepts, LLC, and
Julia M. Phillips, Sandia National Laboratories.

Although the reviewers listed above have provided many constructive comments and suggestions, they were not asked to endorse the views presented at the workshop, nor did they see the final draft of the workshop summary before its release. The review of this workshop summary was overseen by Robert A. Frosch, Harvard University and Woods Hole Oceanographic Institution. Appointed by the NRC, he was responsible for making certain that an independent examination of this workshop summary was carried out in accordance with institutional procedures and that all review comments were carefully considered. Responsibility for the final content of this summary rests entirely with the author and the institution.

Contents

1 INTRODUCTION 1

2 OVERVIEW 3

Workshop Scope, 3

Recurring Themes From the Three Workshop Sessions,

Theme 1. What are the reasons that the Air Force is not adopting directed energy weapon (DEW) systems? Why are the products from Air Force (and other) DEW research and development (R&D) programs not being operationalized for Air Force applications?, 4

Theme 2. What role does the Air Force have in the application of ground-based DEWs to air base defense?, 11

Theme 3. What are potential paths forward for transitioning DEW technology and/or systems to Air Force operations?, 12

3 SYNOPSIS OF WORKSHOP SESSIONS 18

Session 1. February 27-28, 2013, Washington, D.C., 18

Session 2. March 18-19, 2013, Washington, D.C., 30

Session 3. April 24-25, 2013, Washington, D.C., 40

APPENDIXES

A Biographical Sketches of Committee Members 46

B Terms of Reference 40

C Workshop Session Agendas 51

D Workshop Participants 60

ACRONYMS

ACC	Air Combat Command
AFRL	Air Force Research Laboratory
AoA	Analysis of Alternatives
ARDEC	Armaments Research, Development and Engineering Center
C-RAM	counter rocket, artillery, and mortar
CHAMP	Counter-electronics HPM Advanced Missile Project
COCOM	Combatant Command
CONOPS	operational concept(s)
DARPA	Defense Advanced Research Projects Agency
DEW	directed-energy weapon
DIRCM	Directed Infrared Countermeasures
DoD	Department of Defense
DRFM	digital radio frequency memory
EO	electro-optical
HEL	high-energy laser
HEL-JTO	High Energy Laser Joint Technology Office
HELLADS	High Energy Liquid Laser Area Defense System
HPM	high-powered microwave
IED	improvised explosive device
IR	infrared
IRCM	infrared countermeasures
IRST	infrared search and track
JCIDS	Joint Capabilities Integration and Development System
JCTD	Joint Capability Technology Demonstration
JHPSSL	Joint High Power Solid State Laser

LADAR	laser direction and range-finding
LAIRCM	Large Aircraft Infrared Countermeasures
MAJCOM	major command
MDA	Missile Defense Agency
NKCE	non-kinetic counter-electronics
OSD	Office of the Secretary of Defense
R&D	research and development
RDECOM	Research and Development Command
RELI	Robust Electric Laser Initiative
RPA	remotely piloted aircraft
S&T	science and technology

1

Introduction

The U.S. Air Force currently invests significantly in science and technology (S&T) for directed-energy weapon (DEW) systems. Key elements of this S&T investment include high-energy lasers (HELs) and high-power microwaves (HPMs). Other DEW research and development (R&D) efforts include optical beam control for HELs, vulnerability and lethality assessments, and advanced non-conventional and innovative weapons. In 2009, the National Defense University recommended that "Thc DEW S&T community must work to overcome the perception of its unproductive legacy. To do so, the community needs to field systems that will help demystify the new technology. Lower-power systems to defeat soft targets, such as light vehicles and remotely piloted aircraft (RPAs), to counter sensors, and to produce non-lethal effects are suitable for early operational evaluation should be given high priority."[1] This wide range of on-going DEW R&D provides opportunities for the Air Force to consider in transitioning DEW S&T for key capabilities to support the Air Force mission, especially given emerging attention to the Air-Sea Battle concept. In addition, current and future budget limitations facing the Air Force and the Department of Defense (DoD) lend considerable merit to reviewing a range of DEW activities to reduce or eliminate redundant Air Force efforts and leverage on-going and past government-wide efforts where possible.

In this context, the Air Force requested that the Air Force Studies Board (AFSB) of the National Academies' National Research Council (NRC) conduct a workshop to explore the above issues. Members of the AFSB drafted the workshop terms of reference (TOR) in April 2012, and the NRC approved the TOR in July 2012. The NRC then appointed an ad hoc committee to plan and convene a workshop titled "Assessment of Directed Energy Research and Development for U.S. Air Force Applications." Appendix A provides biographical sketches of the workshop committee members. Three 2-day data-gathering sessions were held in 2013 on February 27-28, March 18-19, and April 24-25. Appendix C provides the agendas for the three sessions, and Appendix B lists the workshop TOR. Working with the Air Force workshop topic "champion"—the Office of the Chief Scientist, Air Combat Command—the workshop committee members identified key stakeholders and experts in DEWs to make presentations to the workshop and participate in its discussions.

---

[1]National Defense University. 2009. *Directed Energy Weapons—Are We There Yet? The Future of DEW Systems and the Barriers to Success.* Elihu Zimet and Christopher Mann. May.

In addition to representatives from the Air Force S&T community and major commands (MAJCOMs), the workshop committee invited DEW experts from the U.S. Army, U.S. Navy, Office of the Secretary of Defense (OSD), and the Defense Advanced Research Projects Agency (DARPA) to participate. To learn about the threats that Air Force DEW capabilities might defend against and to assess foreign progress in DEW, the committee invited the Air Force Research Laboratory (AFRL/RD) to provide a briefing on the current threat environment. In addition, a representative from U.S. Pacific Command (PACOM) briefed the workshop participants on current challenges in PACOM's area of responsibility and PACOM perspectives on DEWs. The workshop committee members participated as invited guests in the workshop, as did representatives from the Air Force sponsor. The chair of the workshop committee served as chair/moderator at all three workshop sessions. Most of the presentations and all of the discussions throughout the three sessions of the workshop were conducted at the DoD Secret level. Appendix D lists the participants for each session.

The workshop was conducted under the rules of the NRC for a convening activity, with the purpose of gathering individual views relevant to the workshop topic. In particular, this means that no collective positions—neither views, findings, conclusions, nor recommendations—of the participants were developed or captured for reporting. *All views from the workshop, including any suggestions for future actions by the Air Force or others, expressed in this workshop summary are solely the views of individual participants as understood and interpreted by the rapporteur who authored this summary. Although the chair and other members of the planning committee participated in the workshop, they did so as individuals, and nothing in this report should be construed as a "committee position."* As an aid to readers, the rapporteur has summarized themes and points that were expressed or supported by multiple participants, often across two or three workshop sessions. The content and formulation of these themes and points are the sole responsibility of the rapporteur acting as an observer of the individuals who participated in the workshop and reporting on what participants said. They should not be interpreted as convergence of participants' views or as a consensus—explicit or implied—of the workshop as a whole.

2

Overview

This overview, prepared by the rapporteur, describes several important scoping conditions on the subject matter that was presented and discussed at this workshop but may not be obvious from the TOR. It then describes a set of themes and major points that recurred frequently in the participants' discussions—often across two or all three of the 2-day sessions.

WORKSHOP SCOPE

For this workshop, "directed energy" refers to both coherent light beams produced by lasers and directed beams of microwave radiation. Although the workshop title and the TOR refer to directed energy broadly, the focus throughout the workshop, with the Air Force topic champion's consent, was on directed energy for weapons applications, not lower-power applications for directed energy such as laser designators or lasers used primarily as sensor components (for example, laser direction and range-finding, or LADAR, applications). In short, the workshop participants discussed high-energy laser (HEL) and high-power microwave (HPM) technologies primarily for applications of the energy beam as a weapon. There was some discussion of HEL for laser communications, and some participants noted possibilities for laser DEW systems to have additional specialized roles as sensors, etc.

A second important constraint on scope was that neither offensive nor defensive DEW capabilities for space-based assets were discussed in any of the workshop sessions. Sponsor representatives at the session agreed that neither offensive capability against space-based targets nor defensive capability deployed in space-based assets were within the scope of the workshop.

DEW capabilities of other nations, existing or in development, were addressed at each of the three sessions in at least one presentation as well as in discussions. An important asymmetry, noted multiple times by multiple participants, is that other nations' DEW efforts are largely focused on ground based (or ship-based) weapons for defense against airborne attacks, whereas ground-based defense of U.S. bases or other assets—including Air Force bases—from air attack is not an Air Force responsibility.[2] Although U.S. Army programs for

[2]The foreign airborne HEL systems that were discussed appear to be intended as antisatellite weapons and therefore were outside the scope of the workshop as a potential U.S. Air Force capability.

land-based defensive DEW systems, U.S. Missile Defense Agency (MDA) programs for high-altitude DEW systems, and U.S. Navy programs for ship-based DEW systems were presented and discussed, the emphasis for Air Force applications was on aircraft-based DEW capabilities. To paraphrase a remark made by the chair on this point, "They [other nations] are playing a home [defense] game, we [the U.S. Air Force] are playing an away game, and we have to be able to fly our DEW systems there."

The workshop planning committee provided general guidance on the desired format and content of presentations to the presenters invited for Session 1 and Session 2. For Session 3, individual presenters were asked to address specific topics or questions. The general guidance for Session 1 and Session 2, as well as the presenter-specific guidance for Session 3, are included in Chapter 3.

RECURRING THEMES FROM THE THREE WORKSHOP SESSIONS

A number of themes emerged from individuals' comments and the discussions during the first session and were iterated, with varying amounts of refinement, extension, or revision, by additional individual comments during the second and third sessions. After the first session on February 27-28, the rapporteur drafted a set of emerging themes and discussion points, on which the participants of the second session (March 18-19) were asked to comment. The rapporteur integrated these comments under the draft theme headings and summarized additional recurring themes from the second session. The participants in the third session were asked to comment on this elaborated set of themes and discussion points. After the third session, the rapporteur reviewed all of the captured input and used it as the basis for the following overview.

As noted in the Introduction, this thematic overview, created by the rapporteur, is intended to give readers a condensed view of 6 days' of discussion by a changing set of workshop participants. No consensus, implied or explicit, among the workshop participants is represented here. For some themes, dissenting views and differences of opinion have been included because they help to inform that theme and illustrate the range of views expressed. But a lack of stated disagreement or differences of views on a particular point does not imply there were none. For this overview, silence on any particular point does not imply consent. The order in which themes are presented represents neither a priority ordering nor the temporal order in which they emerged.

Theme 1. What are the reasons that the Air Force is not adopting DEW systems? Why are the products from Air Force (and other) DEW research and development (R&D) programs not being operationalized for Air Force applications?

At the opening of the first session, the chair posed a version of theme 1 as part of his initial comments on the workshop context. During the workshop, several distinct categories of reasons, which are not mutually exclusive, were suggested and discussed:

- The Air Force may not be considering the right DEW applications and concepts of operations (CONOPSs).
- The technology may not be sufficiently mature in aspects that are key to operational capability.
- DEW may not have credibility with the warfighter. The warfighter may not understand or accept the opportunities offered by DEW capabilities.
- The DEW S&T community has perhaps delayed moving from limited demonstration of capability to a practical, operational capability ("kicking the operational can down the road").
- There are challenges (or obstacles) in institutional processes that may have kept the technologically feasible opportunities in DEW from moving to operational capability.

Comments and views voiced by workshop participants on these five general reasons for why DEW R&D results have not resulted in operational capability for the Air Force are described under subordinate themes 1.1 through 1.5, respectively.

Theme 1.1 Is the Air Force considering the right DEW applications and CONOPS?

This theme first appeared during the first presentation of the first session, when Dr. Janet Fender asked if Air Force DEW programs have been addressing the most critical operational challenges for the Air Force and for national defense. In the discussion at the end of that first day, the following points were made:

- Have CONOPSs, which take advantage of the unique capabilities provided by DEWs, been defined for DEWs? "We need to address how DEW capability would be used differently," said one participant, who suggested self-defense for large aircraft (e.g., tankers or AWACS) in a nonpermissive environment as an example of wider CONOPSs than were typically considered. Several others agreed.
- Instead of DEW concepts being touted as a one-for-one replacement for conventional options, one participant suggested, look for other advantages directed energy brings and how to use it in new ways. Again, several other participants affirmed this point.
- With respect to the Counter-electronics HPM Advanced Missile Project (CHAMP), multiple participants across all three sessions saw it as a DEW capability with a clearly defined CONOPSs: nonkinetic counter-electronics (NKCE) of value in operations short of kinetic engagement (e.g., bombing the targeted facilities), where subsequent rapid restoration of a target's electronics capability, as well as denial of attribution, might be operational objectives (Figure 1-1). One constraint on CHAMP that several participants noted was that the number of rounds potentially available through the proposed rapid acquisition program (essentially, refurbished cruise missiles being retired from the inventory) would provide, at most, a raid capability, rather than a campaign-level capability. (CHAMP is discussed further below, under Themes 1.2 and 3.2.)

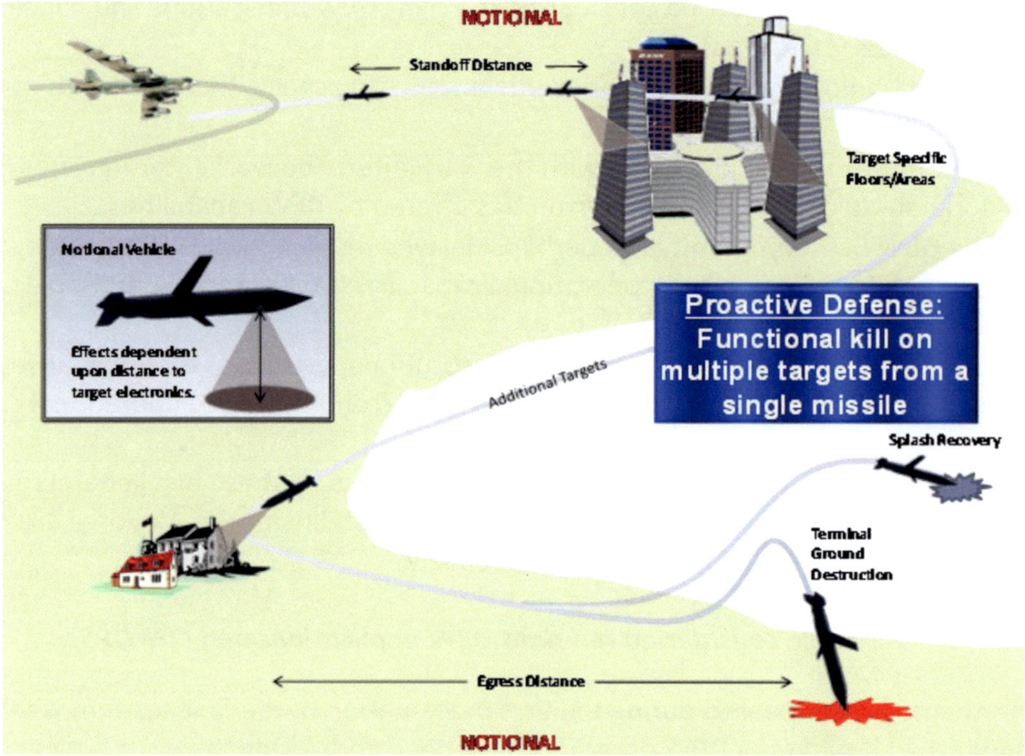

FIGURE 1-1 Notional CONOPS for an NCKE missile such as CHAMP. SOURCE: Robert Peterkin, Chief Scientist, Directed Energy, Air Force Research Laboratory, "Assessment of Directed Energy Research and Development for U.S. Air Force Applications," presentation to the workshop on February 27, 2013.

- A view expressed at the end of Day 1, Session 1, is that the operational community needs to better understand the trade space for DEW versus kinetic solutions (i.e., bullets, bombs, or explosive-delivering missiles). Several participants in later sessions agreed with this point, but there were a range of opinions on what that trade space might in fact look like.
  — For example, a participant commented during the third session review of the first-session themes that operational commanders, specifically Combatant Commanders of the Joint Combat Commands (COCOMs), should identify needs that cannot be met by conventional capabilities [kinetic capabilities], so that DEW would be filling an urgent [unmet] need, rather than replacing a current capability. The same participant also said that there are lower-power HEL missions (e.g., tens of kilowatts) that are needed, but he had not seen, during the three sessions, a "must-do" Air Force mission that requires a high-power laser (100-150 kW) on an aircraft. As further explanation of his point, he said that the size, weight, and power (SWaP) characteristics of HEL systems that are ready today do not support an aircraft-borne high-power laser DEW. In his view, the closest such capability, DARPA's Endurance program, was still at least 5 years from being ready for transition to an operational capability.

— Also during the third session, a different participant said that, as the mission context for the Air Force moves away from environments that are permissive (U.S. air superiority assured), the role of DEW may be to complement kinetic energy weapons—for example, in allowing aircraft to stay on station longer. Due to the potential of deep magazines provided by DEWs, sparser utilization of kinetic energy weapons could be achieved by supplementing them with DEW.

— Starting in Session 1, multiple participants mentioned countering the digital radio frequency memory (DRFM) challenge as important. But while some participants suggested this could be a good early application for an airborne HEL system, others countered that there were alternative approaches to counter-DRFM (see Theme 3.2). As one participant said at the end of Session 3, counter-DRFM is a near-term capability that the Air Force needs, and a laser-based counter-DRFM capability might be possible with today's technology as this capability does not require high power.

- At the end of day 1 of Session 3, a participant said that near-term HEL capability that could be retrofitted on general-purpose aircraft (e.g., fighters) was in the 30 kW range. (Other participants suggested a range from 10 to 50 kW.) At higher power, he said, the platform would have to be for a special mission. He contrasted this situation for HEL DEW with CHAMP, which he saw as substantially changing the probability of mission effectiveness [for certain types of counter-electronics attack]. The same did not seem to this participant to be true for any of the laser applications briefed to that point in the workshop.

Theme 1.2 Is the technology sufficiently mature in aspects that are key to operational capability?

In Sessions 2 and 3, several participants said they had difficulty seeing a near-term or even mid-term HEL application for the Air Force. (See, for example, the closing discussions for Day 1 and Day 2 of Session 2.) They often contrasted the prospects for HEL for Air Force applications with CHAMP as a good near-term prospect for an HPM system and also with ground- and ship-based DEWs for the Army and Navy (see Theme 3.3 discussion).

A technology maturity issue that arose early in the first session, which continued to be brought up during Sessions 2 and 3, concerns the SWaP requirements for HEL systems of sufficient power or energy to add paradigm-changing weapons capability to Air Force aircraft. At the end of Day 1 of Session 1, a participant from Office of the Secretary of Defense (OSD) expressed the view that part of the reason why HEL systems have not been transitioned to operational capability is that DEW programs have tried to take the technology as it was and force it onto an existing platform without doing the SWaP analysis for practical application. This participant suggested that the Air Force continue working on small laser systems suited to aircraft installation, working off what the Navy and Army are doing with applications that allow larger, heavier systems. (See Theme 3.3, which follows up on this suggestion.)

By the end of Session 3, a number of the participants in that session commented favorably on the term-term potential for lower-power (30-50 kW) pod-based laser systems with counter-sensor applications, such as directed infrared countermeasures (DIRCM) and combat ID. (In some comments, the combat ID role seemed to include counter-DRFM.) The DARPA briefing on the Endurance program seemed to convince these participants that a pod-based HEL system for attacking the guidance sensors on electro-optical and/or infrared guided air-to-air missiles (i.e., HEL as an anti-electro-optical/infrared (EO/IR) missile weapon) could meet the SWaP constraints and become operational in the near term (within 5 years). There was extended discussion of what the Air Force could do with a pod-packaged HEL system in this power range, and even as low as 10 kW. Aircraft self defense with a pod was also discussed among the participants.

With respect to challenges for moving HPM technology more rapidly forward, a participant noted during the Session 3 final discussion that the U.S. budget for HPM DEW technology and systems is about $30 million a year, whereas the budget of the People's Republic of China for HPM weapons research is around $300 million.

Theme 1.3 Does DEW lack credibility with the warfighter? Does the warfighter understand and accept the opportunities offered by DEW capabilities?

The views expressed during the workshop on this theme were diverse, and some were in direct conflict with others, even when participants agreed that DEW does lack credibility. Particularly during Session 1, several participants said or implied that a lack of understanding of DEW by Air Force warfighters was an important factor in why DEW technology has not yet achieved operational capability. But several other participants said or implied that the DEW R&D community lacked credibility because of overpromising in the past, either with respect to ultimate technology capability or (more often) with respect to when capability objectives and maturity levels would be achieved. (On the latter, see Theme 1.4.) As the discussions progressed during Session 1, and particularly through Sessions 2 and 3, this direct conflict gave way to more reasoned perspectives that typically brought together aspects of the initial, radically opposed views.

A good example of this evolution is a Session 1 comment from a former warfighter now working in threat intelligence: This is a very complex technological field, he said, and the complexity gives a lot of options, which require a lot of resources [to pursue]. A communications plan is needed to explain the options to the operational community and decision makers. Another former warfighter said that, to move DEW forward, the R&D community needs a "win." To get credibility, the community needs to identify a winning application and get it operational. This view was affirmed in various comments by a number of participants from different roles and backgrounds during Sessions 2 and 3, including technologists and technology systems managers. One participant added the caveat that "a win is needed in an area that needs a win. Do not try to do something [with DEW] that kinetics already does well enough."

Diversity in views continued throughout the workshop on the best way to establish credibility by delivering a win, although the suggestions were typically not contradictory or mutually exclusive. During Session 1, an early suggestion was that, for advanced concepts, it may be better to aim explicitly at a capability demonstration as the objective, rather than an operational capability as the outcome. The point of this suggestion was to avoid the appearance of overpromising. But many participants thought that getting beyond even capability demonstrations to a prototype, which would be closer to an operational capability, was both needed (for a win) and feasible. (For more on prototypes as the objective of R&D efforts, see Theme 3.1.) The missiles fabricated for the CHAMP Joint Concept Demonstration (JCT) ground and flight testing were noted by several participants as a good example of what they meant by a prototype. And the number of comments that favored aiming for a prototype increased during Sessions 2 and 3. Whereas the chair had agreed in Session 1 that DEW R&D programs may not always lead to operational capability, by Session 3 he was pressing more strongly for getting to a prototype: "Unless we partner the science with integration engineering to make a prototype, we will never lead to a fielded system."

There continued to be recognition, expressed in several ways by different participants, that educating and informing warfighters and decision makers about DEW opportunities and challenges would be important and probably essential. In the discussion at the end of Day 1 of Session 3, a participant said that there is still a resistance [in the Air Force culture] to changing the fundamental phenomenology of Air Force weapons. Several views were offered on ways to change the perception of DEW. One line of argument suggested by a participant was that, since other countries are fielding HEL weapon systems and moving out on DEW systems (for example, the Russians have an airborne laser that they say will go operational), should not the Air Force be moving forward with DEW capabilities for its missions? The participant noted that the U.S. intelligence community is undertaking a major study of foreign DEW capabilities and R&D activities. However, other participants questioned the force of that argument, noting the asymmetry between the "home-field advantage" of large-SWaP HEL systems for ground or ship-board defense compared with the Air Force's interest in systems suitable for installing in aircraft.

Theme 1.4 Has the DEW community been, in some sense, kicking the can down the road, in a way that has delayed transitioning technology to operational systems?

The participants in the Day 2 wrap-up discussion at the close of Session 3 returned to this issue, which had been raised initially during Dr. Fender's presentation early in Session 1 and was discussed again during the Day 1 wrap-up discussion. Several participants, at different times during the workshop, voiced the view that the DEW R&D community has tended to go from a fairly successful concept demonstration to working on a more promising solution that was further down the road, pursuing advances in the underlying technology rather than focusing on getting to an operational capability, even if limited. During Session 1, one participant expressed this view as a pattern of industry [e.g., military contractors] doing a great demonstration, then starting over from a new requirement set. A somewhat different take, from a DEW technologist, was that laboratories like to work on exciting, new stuff, but dealing with cost and other issues for operationalizing a system are not as exciting.

This perspective was often tied in with suggestions that the R&D community needs to focus now on getting to prototypes, not just concept demonstrations (see Theme 3.1 for more on this line of thought). CHAMP was mentioned several times by different participants as a good example of how to proceed toward a prototype that the warfighter community could see as an operational capability.

During the Session 3 wrap-up, a somewhat different rationale for delays by the R&D community was suggested. A participant asked if part of the problem might be risk aversion over testing and failing a test, rather than experimenting, implementing an initial limited capability, and moving forward through incremental improvements to that capability. Again, discussion of this thought led to comments on the value of trying to get to prototypes and not just technology or concept demonstrations. Finally, the workshop participants also discussed at the end of Session 3 the possible resistance to DEWs from vested interests (for example, current missile and gun manufacturers).

Theme 1.5. Are process improvements needed to enable transition of mature DEW technology into operational capabilities?

This theme originated in a question the chair posed to the Session 1 participants during the wrap-up discussion on Day 2: "What are the three most important process improvements to allow transition of mature capability?" Several comments at that time related to improving understanding in the user (warfighter) community of DEW potential capabilities and development status and challenges. For example, one participant said the R&D community needs to get the message out to set the stage for the need for DEW; he stressed the importance of communicating an integrated picture. A participant with an operational and military intelligence background suggested that the community needs to communicate to operators in language they understand—lay out the vector for where the technology advocate wants to go with a capability and then deliver on it.

A different take on process improvement is represented in a Session 2 comment from the chair on this theme. He said that it does not appear that the existing coordinating boards and processes that should be relevant to directing and transitioning DEW R&D activities are being used or have been absorbed in other priorities. During Session 3, he again said that the Air Force and DoD do have processes to get warfighters, AFRL, and DARPA together to discuss options, but it appeared to him that these processes are not being exercised. The processes that would help in transitioning DEW concepts into operational capability in a reasoned way are not being followed. Several other participants agreed on the importance of getting the R&D community (AFRL and others), military contractors with DEW expertise, and Air Force operational leaders together. One participant said that the Air Force's Flagship Capabilities Council (FCC) has all the right people on it; the general officers on the FCC could make decisions to move forward toward transiting DEW. He agreed with the chair that the right processes exist, but they are not happening as intended.

Several participants, at different times, called for a capabilities-based assessment that would include DEW options for critical Air Force capabilities along with conventional kinetic alternatives. One former warfighter and acquisition manager who voiced this need for a capabilities-based assessment added that the technical, tactical, and financial values of DEW options need to be demonstrated. The Air Force used to have a process for setting S&T priorities, he said, and that process needs to be revived. A number of other participants agreed with this view. During Session 3, a participant similarly noted that DOD is not doing the same kind of gap analysis studies that used to be done.

Digging deeper into the issue of current priority-setting processes and how well they work, some comments implied that the Joint Capabilities Integration and Development System (JCIDS) process was an impediment to transitioning good DEW concepts, and some other approach should be sought. But other participants countered that JCIDS was not going away, and DEW advocates had to learn to work the process—for example, by identifying mission requirements that reasonably mature DEW concepts (not blue-sky future possibilities) alone could meet or that were unequivocally superior to conventional alternatives. One participant saw a breakdown in the requirements process relevant to this technology area; another agreed that there were process breakdowns, but he also faulted cultural issues within the Air Force and DOD such as conflicts with established equities, comfort levels in the organizations, and lack of understanding of directed energy, all of which interfere with adopting disruptive technologies like DEW.

Other participants noted that the legislation that created the High Energy Laser Joint Technology Office (HEL-JTO) also created a High Energy Technology Council, chaired by a senior OSD official, to provide policy-level oversight of the HEL-JTO. He and several other participants opined that this council seems to be taking a narrower view of the scope of the HEL-JTO than the authorizing legislation called for. What the role of the HEL-JTO should be was a recurring issue that also pertains to Theme 3.1. After one of the discussions of the appropriate role for the HEL-JTO, a participant said the larger point is that the limited amount of funding available for DEW work needs to be spent more wisely. That will take leadership, he added. Several participants thought that the leadership needed should come from OSD or a DoD-wide entity. Others thought that support from several Combatant Commanders could provide sufficient mission pull.

Theme 2. What role does the Air Force have in the application of ground-based DEWs to air base defense?

During Session 1 of the workshop, several participants, including the chair, were favorably impressed by progress at the U.S. Army Space and Missile Defense Command (SMDC) on a ground-based solid-state HEL system as a defensive weapon to protect bases from incoming rocket, artillery, and mortar rounds (counter-rocket, artillery and mortar, or C-RAM). During that session, several participants suggested that air base defense might be a good early application of HEL for the Air Force. By the beginning of Session 2, however, the chair stated clearly that ground-based defense of U.S bases or other overseas military assets was a legally mandated Army mission, not an Air Force mission.

Many participants accepted this statutory reality, but the Air Force's interest in and potential responsibility for defense of overseas air bases recurred several times as a topic of extended discussion during both Sessions 2 and 3.[3] Although none of the other participants expressed disagreement with or qualified the chair's comment that "the Air Force has a vital interest in [overseas] air base defense," there were a number of opinions expressed on how it might pursue this vital interest. A number of participants agreed that the Air Force would need to be more active in ensuring adequate capability for air base defense than just following what the Army was doing; some stronger form of collaboration would be needed. Some participants thought that Air Force funding support for Army work on ground-based HEL, as a defense

[3]See, for example, the synopsis of the discussion on air base defense during the Day 1 wrap-up discussion for Session 2.

against a comprehensive set of airborne threats, was worth considering to ensure that air bases could be defended. Others strongly disagreed, arguing that providing funding support for what was an assigned Army mission would just further drain Air Force resources needed for other priorities. Still other participants thought the Air Force may need to do development and system transition of its own, if the Army's base defense systems were not demonstrably successful against threats such as cruise missiles, longer-range ballistic missiles, aircraft, and potential threats such as swarms of armed RPAs. In the view of one participant, the Air Force should be as focused on air base defense for self preservation as the Navy is on protecting aircraft carriers.

Day 1 of Session 3 began with a presentation on Air-Sea Battle CONOPS by representatives from the Air-Sea Battle Office in HAF/A5X. The workshop planning committee prepared guidance for this presentation, which was intended to clarify the issues about overseas air base defense that participants had raised and discussed in Sessions 1 and 2. However, the presentation did little to resolve the fundamental question of the best way for the Air Force to pursue this "vital interest."

Theme 3. What are potential paths forward for transitioning DEW technology and/or systems to Air Force operations?

As one might guess, the participants' various suggestions for a path forward to transition DEW R&D results to an operational capability depended in part on which of the answers to the Theme 1 questions seemed most relevant to an individual participant. The following four subthemes are discussed below:

- What R&D is needed to move DEW technology forward?
- Which technically mature systems could become fielded operational capability in the near term?
- How might HEL applications for the Air Force be leveraged from HEL development programs of the Army and Navy?
- Are there particular topics or issues related to Air Force DEW applications that would benefit from a further exploration through a targeted workshop or study?

Theme 3.1 What R&D is needed to move DEW technology forward?

During Day 1 of Session 1, the participants were briefed on DEW R&D initiatives at AFRL, the HEL-JTO, the Army SMDC, and DARPA. In the discussion at the end of that day, several participants highlighted what they saw as common R&D issues that could be addressed collaboratively to move forward faster and with greater cost-effectiveness. A comment that was supported by a number of participants was that HEL improvements in such areas as power storage, cooling (thermal management), and solid-state diode efficiency could help "across the board" and could influence how HEL capability looks to decision makers outside the R&D community.

While a number of participants saw work on these common technology enablers as a good role for the HEL-JTO, several participants supported a comment, made during Session 3, that the HEL-JTO has not been taking on some technology development that, under its charter, it should be doing. Are the resources being spent well on technology development, this participant asked, or should the HEL-JTO be focusing more on technology demonstrations? Another participant thought the HEL-JTO work on the Joint High Power Solid State Laser (JHPSSL) and the Robust Electric Laser Initiative (RELI) had been valuable in moving the HEL field forward, but too much of the rest of the HEL-JTO portfolio was spread around on a large number of small efforts that collectively did not have much impact.

A comment relevant to Theme 1.5 was that, although the HEL-JTO was meant to be an organization strong enough to bring the military services and DARPA together to make progress on HEL concepts, it has not succeeded in getting DARPA to participate actively. Another participant suggested that the OSD High Energy Council, which is supposed to provide oversight of the HEL-JTO, could exercise authority over the services and DARPA to strengthen this cross-agency and cross-service role.

During the Session 3 discussions, the HEL-JTO was mentioned in conjunction with moving DEW technology beyond demonstrations and into prototypes. As discussed under Theme 1.1, many participants suggested that aiming for prototypes of DEW systems would help get a “win” that could engage greater interest from the operational community. For example, one Session 3 comment was that militarily useful and suitable prototypes are a reasonable next step for DEW. This participant, like a number of others after the DARPA presentation, thought that a pod-based HEL system that could be fitted to selected aircraft would be a good candidate for a prototype. Another participant thought that the operational-capability goal for DEW prototypes should come from the MAJCOMs, not the laboratory.

A number of participants pointed to CHAMP as a good model for moving to a prototype. For example, one Session 3 participant noted the value of using an existing vehicle/delivery system (the older, conventional cruise missiles that were being removed from inventory) for this first operational capability, to show that the HPM technology could be packaged in a system and used. Because the CHAMP “prototype” had been developed through a JCTD, a number of participants in Session 3 discussed whether JCTDs are a good way to effect transition—there were both pros and cons expressed. For some participants, prototype systems were seen as a way to focus R&D on integration and system engineering. During Session 3, for example, the chair said he had been convinced in the course of the workshop that technical progress is being made on both HPM and HEL, but, he added, the DEW field is lacking the integration and packaging efforts to get to prototypes. Another participant, agreed, saying that a prototype means a packaged, integrated system. Yet another participant stated that the Air Force has only S&T funds for DEW, while the Army and Navy have both S&T and acquisition funds for DEW. The same participant noted that the Air Force does not have funds to transition DEW; this impacts the development of prototypes.

During all three workshop sessions, additional drivers for R&D were discussed by various workshop participants and included improvements in SWaP and employment options.

Theme 3.2 Which technically mature systems could become fielded operational capability in the near term?

Another way that views on potential paths forward were expressed was in response to a question posed by the chair at the end of the second session: "What are the three technically mature DEW systems that can go into the field in the near term?" Over the three sessions, many participants said that CHAMP was technically mature, had a clear although limited mission role for which there were not existing alternatives, and had a potential acquisition path in place that was both technically and fiscally feasible.

To many of the participants, potential HEL weapons for airborne DEW seemed likely to be further out in time than the HPM capability represented by CHAMP (see Figure 2-1).

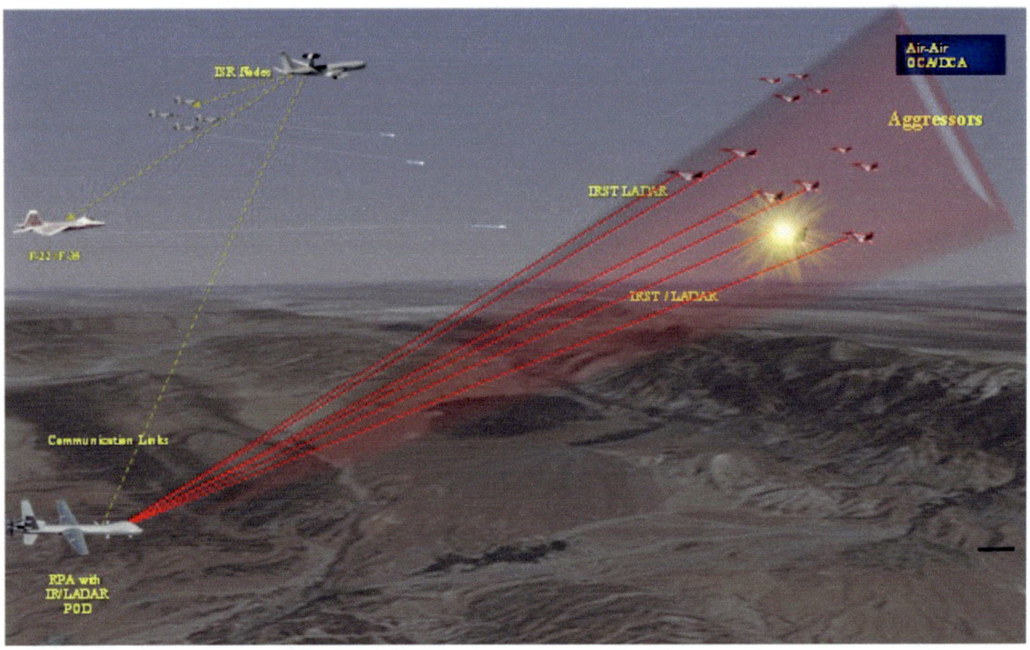

FIGURE 2-1 Notional roles for airborne HEL weapons: offensive counter-air (OCA) and defensive counter-air (DCA). On the left is a laser communications role. On the right is a a counter-infrared search and track (IRST) role. SOURCE: David Robie, Office of the Chief Scientist, Air Combat Command. "ACC DE Enabling Concept," presentation to the workshop on February 27, 2013.

The SWaP constraints for a HEL weapon system that could be fitted to either legacy or next-generation fighter aircraft led many participants to look at HEL systems in the lower-power range (10 to 50 kW) as the feasible option for near- or mid-term roles. As noted above, a number of participants saw a pod-based system, building on the technology represented in the DARPA Endurance program, as the most likely near-tem prospect, and several potential roles for such a system were suggested and discussed.

Early in the workshop, during Session 1, the challenge of defeating DRFMs being used to spoof the radar reflection from an adversary aircraft became a role that a number of participants thought could be a win for a HEL system. While some participants thought a weapons-class HEL could have distinctive advantages for a secondary, counter-DRFM application, other participants noted that alternative technologies for counter-DRFM include LIDAR and LADAR (the latter using a laser in the sensor-application power range, rather than a weapon-power laser).

By Session 3, there were fewer comments favoring a counter-DRFM role for a pod-based HEL and more in favor of using this class of HEL as an anti-sensor weapon to defeat the EO/IR homing sensors of an adversary's air-to-air missiles. Both advanced Directed Infrared Countermeasure (DIRCM) and Large Aircraft Infrared Countermeasure (LAIRCM) applications were mentioned by multiple participants.

Beginning in Session 1, several participants suggested that laser-based optical communications technology (LaserComms) was a good near-term Air Force application to pursue, as it could be implemented with HEL technologies in the 10 to 50 kW range that currently have SWaP parameters within the range of feasibility for a system on legacy or planned aircraft. Although not a weapons application, LaserComms would, according to various participant comments, give HEL a near-term win (see Themes 1.1 and 1.3), build acceptance and familiarity with HEL systems within the Air Force operational culture, and begin building an infrastructure for operational maintenance and support of a HEL system on fighter-class aircraft. However, other participants noted that LaserComms still faces problems with penetrating cloud cover, heavy precipitation, and other weather conditions, so it would have to be part of a broader communications system architecture that also provided redundant systems using radio frequency communications. Another limitation that some participants noted is that a LaserComms system would only be as robust as the number of communications nodes that had the capability installed. By contrast, these participants added, a pod-based counter EO/IRST HEL weapon could be effective even if it were only installed on a few aircraft (e.g., the first-in sorties).

Theme 3.3 How might HEL applications for the Air Force be leveraged from HEL development programs of the Army and Navy?

As noted under Theme 1.2, multiple participants saw the SWaP constraints of existing DEW technologies, when packaged into a system, as a continuing challenge for weapons systems to be installed in a general-purpose aircraft such as a strike fighter or bomber. By contrast, the current SWaP parameters for HPM and high-power HEL systems were viewed as already being within the range of feasibility for land-based or ship-based weapons with sufficient power to defeat a range of threats. Thus, many participants anticipated that the existing and emerging Army and Navy programs for DEW applications will progress more quickly, with high-power DEW systems suitable for Air Force missions becoming feasible on a longer time frame. This perspective naturally led some participants to ponder how the Air Force might best leverage HEL development programs of the Army and Navy.

A Session 3 participant expressed the rationale for an incremental evolution toward higher-power weapons in the long term in this way: For HEL applications, the Air Force should be looking at lower-power lasers [within the HEL category, i.e., 10 to 50 kW of output beam power]. In the past, the Air Force was focused on megawatt-class lasers, not the lower kilowatt-class. The Air Force needs to move incrementally, this participant suggested, from lower-power applications to higher-power applications.

In several comments on the HEL-JTO, a leveraging approach was mentioned as a reason to encourage that joint office to work on technology issues (power, thermal management, etc.) common not only to multiple Air Force HEL programs but also relevant across the HEL programs of the Army, Navy, and DARPA. A number of participants favored greater Air Force cooperation with the Army and Navy on DEW technology development but opposed commingling funds to do so.

During the Session 3 final wrap-up, a participant said he agreed with others that DEW applications are being pursued everywhere (i.e., by the Army, Navy, and DARPA and by other nations), but the S&T issue for the Air Force is where the Air Force should try to lead and where it should follow and "watch closely."

Theme 3.4 Are there particular topics or issues related to Air Force DEW applications that would benefit from a further exploration through a targeted workshop or study?

After several of the Session 3 presentations, one participant in all three sessions suggested that the Air Force should undertake or sponsor an in-depth study on a few promising DEW alternatives, such as CHAMP. The study scope should include examining countermeasures to the capability options, an Analysis of Alternatives [or a capabilities-based assessment, if the alternatives are not at the AoA stage of maturity], and, in general, do a more in-depth assessment than this current workshop format allows. Such a study, this participant continued, would need to consider both trade-offs among alternatives for a given role and counter-actions to which they might be vulnerable. "We have only heard the positive advocates of DEW," he said, "and no in-depth analysis," which should be done before the Air Force "accepts, develops, and weaponizes a capability."

Another participant suggested that a potential area for a targeted study would be the lower-power class of HEL technologies with potential for advanced active sensor applications, such as emerging solid-state laser technology for an imaging IR missile. Such an active sensor might be of value in a LADAR system to support a counter-DRFM role (also see discussion of counter-DRFM under Theme 3.2). This participant also thought the workshop had not heard enough about counter-sensor applications, which require lower power than do vehicle-shell-defeat applications of HEL. One response to this suggestion for a follow-on study was that much of the work in the counter-sensor area may be at higher levels of classification.

Yet another participant comment at the close of Session 3 was that the workshop had not touched on how DEW might work for counter-ISR applications, particularly high-flying airborne ISR assets. This comment led to discussion of using a weapon-class laser system for a long-range ISR role, as opposed to being a counter-ISR weapon. One participant suggested that, in the far term, the same laser system might be used for both "seeing and touching" an adversary's platform (i.e., as both an active ISR sensor and a defeat weapon).

Another study suggestion during Session 3 was to examine the prospects for pulsed power HEL technology in the Air Force, and not just the continuous-wave (CW) laser technology on which most of this workshop focused. In agreeing with this suggestion, a second participant said that the progress made in CW laser technology should be acknowledged in such a study, with pulsed power examined for what it could add to what CW systems can do. This participant was concerned that a pulsed-power laser study not foster a debate on pulsed power versus CW, which could become yet another reason to further delay going forward with prototypes (kicking the can down the road again). Several other participants agreed.

Pulsed power HEL technology was also discussed during the wrap-up on Day 2 of Session 2. One participant said that, from a warfighter (COCOM) perspective, pulsed power was interesting for a counter EO/IR missile role. Another participant added that pulsed lasers may be the only effective HEL countermeasure to the more advanced class of adversary surface-to-air missiles ("double-digit SAMs").

A participant noted that, in the HPM area, the workshop had not addressed much other than CHAMP (i.e., the NKCE role for HPM). Another participant agreed and added that the Robert Peterkin briefing during Session 1 had included more than just the counter-electronics application of HPM, but the Air Force is looking for HPM applications that will fit on an aircraft. This participant said that the ground-based MAXPOWER demonstration was reasonably successful for countering improvised explosive devices (counter-improvised explosive device), but that is not an Air Force mission.

During the Session 3 final wrap-up, while the effectiveness of the HEL-JTO was being discussed, a participant suggested that a more complete review was needed of what the HEL-JTO should be doing; that is, what its role should be.

3

Synopsis of Workshop Sessions

SESSION 1
FEBRUARY 27-28, 2013
WASHINGTON, D.C.

For Session 1, the workshop planning committee sent all presenters the following guidance on the desired format and content of presentations:

> Tables of potentially rewarding high-energy laser (HEL) and high-power microwave (HPM) technologies and systems that are being developed that could be relevant to the Air Force, particularly for defensive applications. For each technology and system, a brief description and its expected performance. Please also provide the period (time line) of R&D [research and development] and testing and when a decision about deployment might occur. If available, please also provide the suggested number of systems that would need to be acquired and the cost of such a system (including development costs).

Wednesday, February 27, 2013

Welcome and Introductions
Gen Ronald E. Keys (USAF, Ret.), Session Chair

Gen Ronald Keys opened the meeting with an overview of the terms of reference for the workshop (Appendix B). He suggested that the workshop objective be viewed as providing useful input to the Air Force on the following key questions:

- What is the root cause, or causes, for why directed-energy weapon (DEW) systems are not being adopted for Air Force applications? Why are the products from R&D programs not being operationalized? Is it, for example, because:
  - The technology is not yet ready;
  - The DEW alternatives are not sufficiently better than other options (e.g., projectiles); and/or the DEW alternatives are not competitively affordable (including the cost of engineering them into current platforms, etc.)?
- Does the technology work in the "mud, blood, and beer" of real-world operations, not just in the laboratory?

- What kinds of issues need to be addressed to facilitate adoption?
  — Tactical issues—the technology works but is not tactically useful (superior to alternatives)
  — Cost issues
  — Intellectual acceptance issues—operational practitioners do not understand the benefits, do not see DEW solutions as a better way of operating

In Gen Keys's opinion, DEW technology has to win its way onto the battlefield. It must add value by doing something better than non-DEW options can, or something that they cannot, do. It has to have a reasonable cost structure and offer affordable capability. Proponents must answer the questions: "Where does it fit?" and "What Department of Defense (DoD) capability gap does it fix?"

Air Force Champion's Vision for Directed Energy
Dr. Janet S Fender, Chief Scientist, Air Combat Command (ACC)
Dr. David Robie, Office of the Chief Scientist, ACC

Dr. Fender and Dr. Robie prepared abstracts of their presentations for inclusion in this report, which have been Incorporated below as Box 3-1 and Box 3-2, respectively.

Rapporteur Comments: In addition to the points made in Dr. Fender's abstract above, her presentation and the interactions during and after it with other workshop participants raised the following points that are of particular relevance for themes that emerged during this session and the workshop as a whole (see Chapter 2).

- Have the Air Force and the DEW R&D community taken advantage of the unique advantages and capabilities of directed energy—not viewing it as simply a bullet-replacement?
- Have Air Force programs been addressing the most critical operational challenges and national challenges? For example, aircraft self-defense, although important in the long term, may not be the biggest national/Air Force challenge in the near term.
  — Dr. Fender and several other participants suggested that DEW technology development aim at high-priority missions.
  — Another suggestion, made by Gen Keys and seconded by several other participants' comments, was that the R&D community ought to seek functional requirements for which DEW systems are best option. For example, do not seek a general requirement to integrate [high-energy] lasers into aircraft. Rather, according to this line of thought, the community would do better by making the case for DEW as a solution for an important functional requirement.
- Dr. Fender's point that successful demonstrations have been viewed as stepping stones to a next generation of technology advancement, rather than leading to an operational capability, evoked many comments and related discussion. A number of participants agreed with one participant's response that the DEW R&D community needs to get products operational, not just "continue kicking the can down the road

on when useful capability will be ready." (Related comments and discussion recurred frequently enough throughout the workshop to make this an important theme; see Theme 1.4.) Several other participants thought there could be strong "user pull" for DEW systems, if it were shown to provide compelling solutions [for Air Force capability gaps or emerging threats]. But the solutions have to be affordable, and there must be a credible maturation plan for them.

BOX 3-1

Directed Energy: An ACC Perspective
Dr. Janet S Fender, Chief Scientist, ACC

Airborne DEWs and adjuncts offer potentially significant asymmetric advantages as well as opportunities to realize innovative concepts, including deep magazine weapons, electronic warfare, novel imaging, and secure high-bandwidth communications. Although the Air Force has identified gaps optimally filled by DEWs, executable technology-maturity plans lag tech need dates for integration into future systems. Furthermore, investment strategy is complicated by the lack of a top-level Air Force perspective of highest-priority needs best met using DEWs. Exquisite lethal and nonlethal capabilities of laser and high-power microwave technologies have been demonstrated in realistic environments since the mid-1970s. However, each resounding successful demonstration was viewed as a stepping stone to a next-generation system with more advanced technology rather than an initial capability for operations. Therefore, the preponderance of Air Force investment in DEWs remains in exploratory and advanced research and development rather than technology maturity. Additionally, mutually beneficial partnerships among services and DARPA have been difficult to develop. Now in an era of austere budgets, a DoD-wide coordinated effort is critical to bring DEW capabilities into operations effectively.

BOX 3-2

ACC Directed Energy Enabling Concept
Dr. David Robie, Office the Chief Scientist, ACC

Currently, a number of directed-energy applications will address the most significant mission gaps and needs of the Air Force. These applications include (1) high-power microwaves for counter-electronics applications, (2) high-energy lasers for airbase defense, (3) lasers in concert with electro-optical sensors for target location and identification, (4) laser communications, and (5) high-energy lasers for air superiority missions. Each of these applications is evaluated, assessed, and system-engineered to provide advanced capabilities in a high-threat environment. Finally, each concept is focused on the practical, cost-effective, and technically mature employment of these technologies.

Air Force Research Laboratory (AFRL) Initiatives
Dr. Robert Peterkin, Chief Scientist, Directed Energy Directorate, AFRL

In his introduction, Dr. Peterkin noted that his presentation does not cover AFRL work on Blue Force defense against Red Force DEW threats. The concepts for employment he presented for the near, mid-, and far term are based on developing a fundamental understanding of targets' vulnerability to directed energy. He described how AFRL works from measures of effectiveness for a potential system (objective system), derives subsystem technical performance measures by decomposing the full system into subsystems, and establishes science and technology (S&T) programs to solve problems in achieving those performance measures. Transitioning the AFRL results to products requires working with industry, and AFRL needs help in transitioning results into near-term DEW products.

In the area of narrow-band HPM systems for counter-electronics applications, Dr. Peterkin discussed the Counter-electronics HPM Advanced Missile Project (CHAMP) Joint Concept Technology Demonstration (JCTD) and its SuperChamp follow-on. The key to a long-range counter-electronics application is to concentrate the energy in a narrow frequency range in short pulses. He described how AFRL uses the Joint RF Effective Model (JREM) to optimize firing location and aim-point and to predict probability of kill (pK) for targets. JREM was used in developing CHAMP prior to the system's flight testing. A single CHAMP missile can engage multiple targets and achieve multiple functional kills. In the CHAMP JCTD, the recharge time between engagements was on the order of tens of seconds. Among the concepts studied was the use of multiple engagement passes (fly-bys) to increase the functional effects on the same target. SuperChamp, the follow-on work to the CHAMP JCTD, has a near-term objective of extending the CHAMP counter-electronics capability to operationally relevant ranges. Its mid-term and far-term objectives are to deliver frequency-agile counter-electronics and cyber/electronic effects from greater distances using other platforms.

MAXPOWER is AFRL's high-average-power HPM system for counter-improvised explosive device (IED) use. The objective is to detonate all classes of IEDs at safe ranges from mobile systems, without prior knowledge of IED location. A test version was deployed to theater for a 1-year operational evaluation. Dr. Peterkin also discussed the use of a quasi-continuous wave (CW) millimeter-wave system as a counter-personnel (nonlethal) active denial system (ADS), for use in driving ground-based personnel away from a clear zone around ground-based assets without employing lethal force.

Turning to HEL systems, Dr. Peterkin presented and discussed a conceptual graph of the timeline for technical risk reduction for airborne HEL concepts versus the relative sophistication and specific power of each application (Figure 3-1).

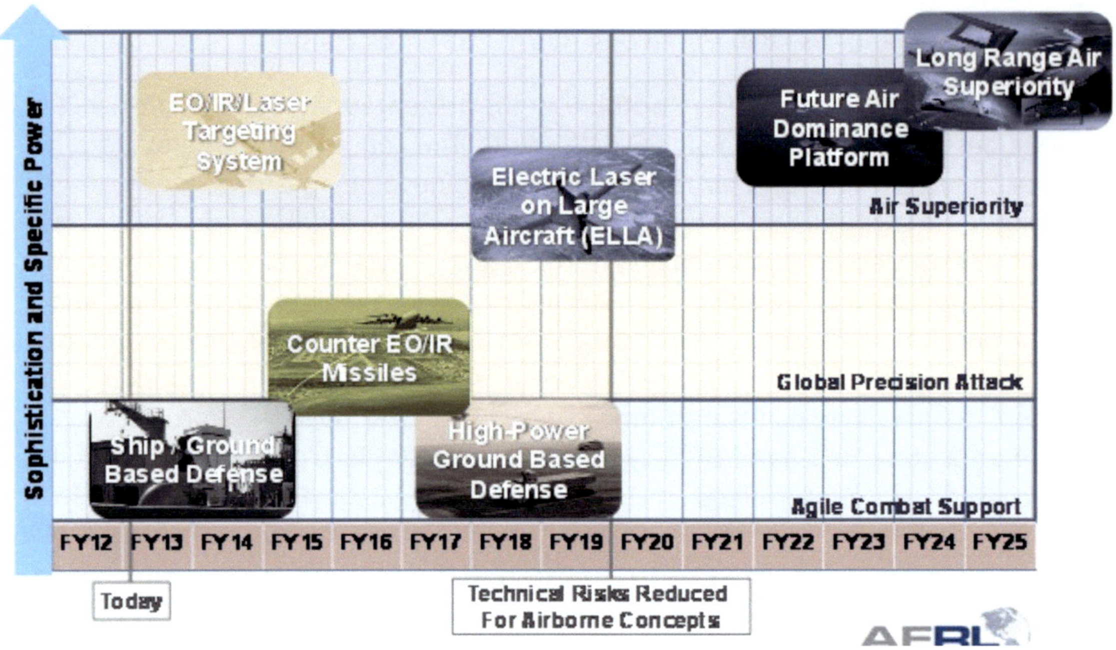

FIGURE 3-1 AFRL laser system concepts and timelines. SOURCE: Robert Peterkin, Chief Scientist, Directed Energy, Air Force Research Laboratory, "Assessment of Directed Energy Research and Development for U.S. Air Force Applications," presentation to the workshop on February 27, 2013.

For "hard kill" laser lethality, the important concept is fluence (energy per unit of area irradiated), rather than power of the emitted beam (kilowatts) or the irradiance (power per unit of area irradiated). Irradiance may be relevant to sensor kill when the laser beam is within the working frequency band of the sensor (in-band sensor kill). Beam quality, spot size, and dwell time are all relevant to pK for hard-kill lethality. Dr. Peterkin discussed how these parameters relate to the challenges for development of antimissile laser defense systems for aircraft and surface ships. Imaging anti-aircraft missiles present a more sophisticated and more difficult-to-counter threat than EO/IR missiles.

AFRL and DARPA are collaborating on a concept for a high-energy electric laser on a large aircraft. This program was enabled by substantial DARPA investment in the HELLADS program. AFRL is also participating in the first four industry efforts of the Robust Electric Laser Initiative (RELI), led by the High Energy Laser Joint Technology Office (HEL-JTO).

High Energy Laser Joint Technology Office Initiatives

Mr. Donald Seeley, Acting Director, HEL-JTO

Mr. Seeley prepared an abstract of his presentation for inclusion in this report; it is incorporated below as Box 3-3.

BOX 3-3

HEL-JTO Initiatives
Mr. Donald Seeley, Acting Director, HEL-JTO

The High Energy Laser Joint Technology Office (HEL-JTO) was established in 2001 as an office within OSD. The mission of the office is to support the designated senior OSD official HEL programs by establishing priorities for the HEL programs, coordinating the programs among the services and agencies, and ensuring that these programs are consistent with the priorities identified in the official's planning and programming activities. In addition, the HEL-JTO manages a portfolio of technology programs to advance the state of the art in HEL weapon system development; maintain the industrial base for HEL-related technologies; and educate current and future scientists, engineers, testers, and warfighters on HEL technologies and systems potential. The HEL-JTO portfolio consists of laser source development, laser beam control technologies, laser lethality sciences, modeling and simulation, satellite and airspace deconfliction, and educational initiatives. Currently, the major initiative for the HEL-JTO is the Robust Electric Laser Initiative, which will dramatically improve the efficiency and beam quality and reduce the size and weight of electrically powered laser devices that can be ruggedized for military applications.

U.S. Army DEW Initiatives at ARDEC

Mr. Charlie Freund, Acting Chief, Directed Energy Branch, U.S. Army Research and Development Command, Armaments Research, Development and Engineering Center

Mr. Freund's presentation provided an overview of DEW efforts at the U.S. Army Armaments Research, Development and Engineering Center (ARDEC). The Sparrow Sentry System for Personnel is an HPM application for screening people at portal control points for personnel-borne IEDs. It resulted from a series of down-scopes to the original concept, which was for a IED route-clearing role. The system was 100 percent effective in testing, but in 3 years of deployment there was never an active detonation (individuals were warned of the screening system's presence). ARDEC is also working on a portable version to screen vehicles at entry control points for IEDs or vehicle bombs and a miniaturized version for screening personnel. Another route-clearing concept in development is the Laser Induced Plasma Channel technology, which works by ionizing air in a channel to an IED target, along which an electrical discharge is sent to detonate the IED from a safe standoff distance.

Solid State Active Denial Technology (SS-ADT) uses a 95 GHz HPM beam to drive personnel away from an area to be protected by heating their skin to an intolerable level. The technology is being considered for both fixed-site and vehicle-mounted applications. SS-ADT has considerable SWaP advantages over the Army's current Active Denial System. A handheld version of the technology is also under development. Funding for the SS-ADT technology is from the Joint Nonlethal Weapons Directorate (JNWD).

U.S. Army DEW Initiatives at the Space and Missile Defense Command (SMDC)

Dr. Kip Kendrick, Chief, Directed Energy Division, U.S. Army Space and Missile Defense Command, Army Forces Strategic Command

Dr. Kendrick described work on a RELI-class HEL system (35 percent laser efficiency) with an advanced beam control system that allows it to fit on a Stryker-class Army vehicle for base defense against rocket, artillery, and mortar fire (counter-RAM). RELI uses a fiber bundle solid-state laser that has the advantage that repairs can be done in the field. The concept for this system was included in the recent Indirect Fires Protection Capabilities II (IFPC II) Analysis of Alternatives (AoA) for the base-defense counter-RAM role. Dr. Kendrick described how this technology has evolved from previous Army involvement in the Tactical High Energy Laser (THEL) and Joint High Power Solid State Laser (JHPSSL) programs. He then discussed the Army's ongoing role in the HEL-JTO's RELI program and the Army roadmap for HEL R&D.

The IFPC II AoA resulted in selection of a missile system even though the HEL system scored well. In light of that choice by the Army Training and Doctrine Command, the workshop participants discussed with Dr. Kendrick his views on the hesitance of the warfighter community to move forward with operationalizing DEW concepts. He emphasized the importance of developing credibility with the warfighter community while development is ongoing. DEW is a paradigm shift for the user, he said, and the S&T community needs to address warfighters' concerns on issues such as supportability and reliability. In response to a question, he said it is important to address doctrine and concept of operations (CONOPS) for a DEW application.

Defense Advanced Research Projects Agency (DARPA) Initiatives

Dr. Richard Bagnell, Strategic Technology Office, DARPA

Dr. Bagnell's presentation described DARPA programs for the High Energy Liquid Laser Area Defense System (HELLADS) and for the Laser Weapon System Module (LWSM) of HELLADS. A key point he made about the LWSM is that the system's footprint is driven by the power and thermal management subsystems necessary to meet the duty cycle and magazine requirements.

Discussion of Day 1 Presentations

Gen Ronald E. Keys (USAF, Ret.), Session Chair

Gen Keys asked the participants to comment on the day's presentations, going in a clockwise direction around the table from his left. After Session 1, the rapporteur organized his notes on this discussion under the following five emerging themes of the workshop:

- Are the right applications and CONOPS for DEW being considered?
- Does DEW have credibility with the warfighter? Does the warfighter understand and accept the opportunities that DEW systems might provide?
- The Air Force DEW R&D community may need to stop kicking the operational capability can down the road.

- What are the potential paths forward for transitioning Air Force DEW science and technology products to operations?
- What are the priority R&D opportunities?

The first three of these themes subsequently became Themes 1.1, 1.3, and 1.4, respectively, in the overview of themes and major points from the workshop as a whole (see Chapter 2). The fourth theme evolved into Theme 3 in the overview, and the fifth theme became Theme 3.1 in the overview. Individual participants' comments related to these five themes are represented in the overview.

Thursday, February 28, 2013

Welcome and Introductions

Gen Ronald E. Keys (USAF, Ret.), Session Chair

In his comments to open the second day of presentations and discussion, Gen Keys offered the following observations:

- DEW proponents should be careful to differentiate between early adopters and beta testers. Based on his experience, Air Force warfighters may have been scared off DEW by too often being used as unwitting beta testers, when they thought they were being early adopters of an operational capability. (See Theme 1.3.)
- It will be important to get the right CONOPS that fit the capability of a DEW technology or system. As an example, he recalled how a HEL system demonstrated for the B-1 bomber and became irrelevant when the CONOPS for the B1 changed from the original design for low altitude and high speed to the current CONOPS of flying at medium to high altitude. (See Theme 1.1.)
- It is okay, he suggested, to present an HEL program as a technology maturation program, but do not sell it as leading to a fieldable weapon system if that is not a likely direct outcome if the proposed program is successful. (Relevant to Themes 1.2 and 1.3.)
- Air base defense is a natural good fit for a big-laser (i.e., large size, weight and power, or SWaP) defense system. Another key application could be anti-access. (In subsequent discussions, particularly during Sessions 2 and 3, Gen Keys emphasized that air base defense is not currently seen by ACC and others as an Air Force mission. See Theme 2.)

OSD Initiatives

Dr. Spiro Lekoudis, Research Directorate, Defense Research and Engineering, OSD

Dr. Lekoudis prepared an abstract of his presentation for inclusion in this report; it is incorporated below as Box 3-4.

BOX 3-4

Topics in Weapons Research and Engineering
Dr. Spiro Lekoudis, Research Directorate, Defense Research and Engineering, OSD

A discussion relevant to aircraft self defense is relevant and appropriate, as technology proliferates and the sophistication of potential adversaries increases with time. Examples of such trends are (1) the challenge to the U.S. superiority in electronic warfare, (2) digital signal processing is expanding, (3) networked passive-active sensor combinations are appearing, (4) software-driven waveform generators are coded, (5) weapon seekers are gaining spectral diversity, and (6) counter-countermeasure capabilities and advanced jamming techniques are evolving. In addition, modern integrated air defense systems are networked, mobile, and redundant. Advanced high-speed air-to-air radar-guided threat missiles, designed to expand current no-escape zones by a factor are also being fielded. As well, electro-optical sensors, immune to some of the existing countermeasures, are being fielded.

Focusing on the potential threat from high-speed missiles, there are two broad areas that both deserve attention: (1) missiles that can sustain high-speed; and (2) missiles that can engage at high speeds. In the first category, the response benefits from accurate tracking, ability to use electro-optical sensing as well as expanding the threat engagement envelop. In the second case, more dangerous because of the reduced reaction time and the lack of help from earlier tracking, the intercept calculations are becoming more complex. It is important to focus attention on the complete kill chain as well as the contribution of new and evolving technological capabilities that can respond to such threats. Recent demonstrations of engagement with very compressed timelines and the increased power available to directed energy systems are two examples of defense science and technology that could contribute to future challenges of aircraft self defense.

Department of Energy (DOE) Initiatives
Dr. Vance Behr, Deputy Director for Military Systems and Technologies, Integrated Military Systems Center, Sandia National Laboratories

Dr. Behr prepared an abstract of his presentation for inclusion in this report; it is incorporated below as Box 3-5.

BOX 3-5

Sandia National Laboratories' Directed Energy Initiatives
Dr. Vance Behr, Deputy Director, Military Systems and Technologies, Sandia National Laboratories

This talk presented information on a variety of directed energy initiatives that are underway or were recently conducted at Sandia National Laboratories. The topics were broadly divided into those dealing with Ultra Short Pulse Lasers (USPL), HPM, and supporting technologies. In the area of USPL, a number of topics were reviewed, including defensive applications of USPL, a few results from the recent Military Utility Study completed for HEL-JTO, and the different effects of USPL versus continuous wave or long-pulse lasers in various media. A number of activities were reviewed in HPM application space, including the Sandia pulsed power system contribution to the recent CHAMP JCTD, efforts in electronic battle damage assessment, efforts to model the coupling of high-power transient electrical waveforms into electronics, design efforts for frequency agile and steerable HPM antennas, compact HPM sources, and effects measurement of 8-95 GHz radio frequency illumination of representative unmanned aerial systems flight status and command systems. Research in the area of photoconductive semiconductor switches was also presented because of its potential for use in HPM energy delivery systems.

U.S. Missile Defense Agency (MDA) Initiatives
Dr. Steve Post, Laser Technology Program Manager, MDA

Dr. Post began by reviewing the chronology of the Airborne Laser (ABL) program. He noted that the ABL system was incredibly complex, in part because the chemical-oxygen-iodine laser (COIL) required a lot of safety systems on the aircraft. Among the lessons learned from this demonstration program, he noted that it successfully engaged both solid and liquid boosters representing long-range ballistic missile targets. In his opinion, the ABL laser system would work for area defense. The system was sold as ready for transitioning to an operational capability ("ready for prime time"), but in fact most of the technology subsystems were one-of-a-kind integrations. Next, Dr. Post described MDA's continuing program in DEW defensive systems, including the Diode Pumped Alkaline Laser System (DPALS) and systems carried on remotely piloted aircraft (RPA), and the Fiber Combining Laser (FCL) program. In response to a question on MDA collaboration with the HEL-JTO, Dr. Post said there were informal synergies across programs, rather than formal collaborations.

Lincoln Laboratory Initiatives
Dr. Daniel Ripin, Assistant Leader, Laser Technology and Applications Group, MIT Lincoln Laboratory

Dr. Ripin described the status of the technology for combined fiber lasers, including advances in beam-combining technologies and quantum cascade lasers. Lincoln Laboratory is investigating hybrid beam-combining techniques (using both wavelength beam combining and coherent beam combining), which could be used to combine a hundred to a thousand single fiber lasers. Dr. Ripin sees fiber amplifier evolution, as in the DARPA goal to produce a 3 kW single-fiber laser, as a good example of laboratory-industry partnering. The DARPA program is currently achieving 5 kg/kW for the Excalibur laser system mounted on a Reaper RPA. Excalibur uses eight 3-kW fibers in the laser array.

Office of Studies and Analyses, Assessments and Lessons Learned, Headquarters U.S. Air Force
Directed Energy Implications for Future Force Structure
Mr. Jeffrey Saling, Chief, Mission Analyses Division

This presentation reviewed recent force structure analyses by the Office of Studies and Analyses, Assessments and Lessons Learned, Headquarters U.S. Air Force, including the 2009 Directed Energy Net Assessment and the 2013 Directed Energy Force Application Study. The purpose of the 2009 assessment was to study in depth the effects of projected adversary directed energy air defenses on U.S. air combat missions. The key question it sought to answer was "How will directed energy challenge or marginalize our advantage in air and space?" The analysis indicated that anticipated Red Force defensive DEW capabilities, including both HPM and HEL systems, could substantially compromise the effectiveness of Blue Force air attacks. From the modeling and simulation of representative scenarios, the conclusion was that a Red Force integrated air defense system infused with DEW defensive anti-air systems poses a significant risk to Blue Force strike mission success, but it was difficult to fully assess the impact on Blue Force systems without additional susceptibility testing to feed into the models.

The second part of the briefing reviewed a 2011 presentation that Mr. Saling and Dr. Robie had prepared on Blue Force DEW systems as an enabling concept for future mission scenarios. The DEW concepts that were included in that presentation were long-range air superiority (LRAS) using DEW systems for both offensive counter-air and defensive counter-air roles, a next-generation tactical laser weapon (assuming a 150 kW solid-state laser) for nonkinetic attack with low potential for collateral damage, combat ID using a synthetic aperture LADAR employed as a cued sensor, nonkinetic counter electronics (a CHAMP-like air-launched cruise missile), and a Laser Strike Fighter concept for both offensive (engaging tactical air targets and ground targets) and defensive (counter-SAM and counter air-to-air missile) roles.

The third part of the briefing was the 2013 Directed Energy Force Application Study, which ACC had requested. This study included aircraft-based DEW concepts for offensive and defensive counter-air in air mission scenarios run with the Brawler scenario simulation system. The concepts studied included LRAS with a large (1.5 MW) COIL mounted in a wide-body aircraft and a Laser Strike Fighter configured as an F-35 B equipped with a 150 kW solid-state laser. The defensive counter-air scenario with the LRAS concept showed increased Blue Force survivability and increased Red Force attrition compared with the no-LRAS scenario. Similarly, the longer-term Laser Strike Fighter concept showed an increase in Blue Force survivability and an increase in Red Force attrition relative to the same counter-air scenario run with a Joint Strike Fighter without the HEL weapon system.

Joint Non-Lethal Weapons Directorate (JNLWD)
Ms. Susan Levine, Principal Deputy Director (Policy and Strategy), JNLWD

Ms. Levine said that, although JNLWD is heavily invested in directed-energy options, the nonlethal weapons for Air Force roles are limited to applications for air base security forces. She reviewed the JNLWD program management structure and the DoD-wide and Air Force-specific budgets for nonlethal weapons. She discussed the Joint Non-Lethal Effects capability gaps and the top tasks included in the counter-personnel and counter-materiel Initial Capability Documents. Ms. Levine discussed active denial systems and roles with the workshop participants, including HPM and HEL applications like those discussed by Mr. Freund for crowd dispersion, boundary protection, stopping ground vehicles, and stopping small boats.

Discussion of Day 2 Presentations and Planning for Session 2
Gen Ronald E. Keys (USAF, Ret.), Session Chair

For this wrap-up discussion of the first session, Gen Keys suggested that the participants address the following four questions:

- What are the three DEW mission areas for the Air Force to play in?
- What are the three technically mature systems that can go into the field in the near term?
- What are the three most important process improvements to allow transition of mature capability?
- What are the three critical challenges or fatal flaws that could kill success?

After Session 1, the rapporteur aligned the individual participant's comments under these four headings. In many cases, the comments did not directly answer one of the four questions posed but were in some way relevant to the topic suggested by the question. In preparing the workshop overview (Chapter 2), the first question and comments related to it were integrated with Theme 1.1. The second question and comments related to it became Theme 3.2. The third question was integrated with Theme 1.5. Comments related to the fourth question were integrated with comments relevant to one of the five subthemes under Theme 1.

SESSION 2
MARCH 18-19, 2013,
WASHINGTON D.C.

Monday, March 18, 2013

Welcome and Introductions
Gen Ronald E. Keys (USAF, Ret.), Session Chair

Gen Keys opened the meeting and described a meeting he had with ACC staff, including Gen Hostage, ACC commander, after the first workshop session. A key point made by Gen Hostage was that air base defense is mandated by U.S. statute to be an Army responsibility. Gen Keys discussed with the workshop participants whether this legal stricture would apply to all airborne threats to an overseas Air Force base, as well as to protection from threats "just outside the fence," which has traditionally been the kind of defense provided by the Army for overseas air bases. Various participants mentioned cruise missiles, longer range ballistic missiles, and RPAs as potential future threats to Air Force bases that might be beyond the capability of existing or planned Army base defense systems.

With respect to the status of CHAMP, Gen Keys said that an important issue for real-world CONOPS was whether a non-kinetic counter-electronics (NKCE) missile such as CHAMP would be in the inventory for raid-type operations versus campaign-level operations. Even the maximum number of CHAMP missiles that potentially could be produced from the available conventional cruise missiles being removed from the Air Force inventory would only provide enough NKCE missiles for raid-like operations. For the CHAMP rapid acquisition program to proceed, he said, one or more combatant commanders would have to make it priority, which would mean giving up something else on their priority list. Dr. Janet Fender, ACC chief scientist, who participated in the morning discussions by telephone, agreed with Gen Keys's account of his ACC meeting and added that Gen Hostage had said CHAMP would have to be a priority for a COCOM for ACC to proceed with either of the production options.

Discussion of Key Themes and Points from Session 1
Dr. Robert Katt, Rapporteur

Dr. Katt reviewed with the workshop participants a draft of highlights (key themes and points made by individual participants) from the first workshop session on February 27-28. Gen Keys and other participants offered comments, which were captured for incorporation with the participants' comments from the first session. Among the new comments were the following, which are particularly relevant to the content of the overview (Chapter 2).

- Gen Keys expanded on his rationale for his initial suggestion that the workshop should try to shed light on why DEW systems are not yet being used by the Air Force. He suggested that participants try to address the status of directed-energy capabilities in development, what might be needed for them to win acceptance as operational systems, and how to address the cost issues associated with moving to an operational capability. What are the issues for practical use of DEW by the Air Force?

- With respect to Red Force use of DEW systems, Gen Keys noted that an adversary would likely have an asymmetric advantage in that large, heavy DEW systems can protect fixed bases, whereas the U.S. Air Force has to address difficult SWaP issues to have DEW systems that are practical for use on aircraft. "Directed-energy weapons are an away game for us, a home advantage for the other side," he said.
- Several participants offered suggestions on how to move DEW development forward to operational capability for some application(s). Support from an Air Force Major Command (MAJCOM) or a Joint Combatant Command (COCOM) was mentioned several times. Maj Gen David Scott, a member of the workshop planning committee, suggested that a good answer would be needed when a Combat Commander asks, "What is DEW going to do to help me win the fight?"

Barriers to Transitioning Technologies for Operational Environment
Mr. Terry Jaggers, Director, Air Force Studies Board, National Research Council; Former Deputy Assistant Director of the Air Force for Science, Technology, and Engineering

Mr. Jaggers began his presentation by reminding the participants that the aim of the workshop is to gather ideas for the Air Force to consider, not to advocate which particular applications should go forward. Based on his previous position in Air Force acquisition, he described the kind of environment and appropriate drivers to support the transition of a DEW technology to a fielded capability. Successful transition has occurred in the past, he said, when (1) scientists and engineers understand the operational environment, (2) program managers can see a balance between high- and low-risk options with the laboratories and industry partners, and (3) warfighters are involved from the start and throughout in the concept, system design, and test.

Several other participants commented on issues with getting warfighter involvement to the degree Mr. Jaggers suggested would be needed. Again, several participants favored involvement and backing from one or more COCOMs. The importance of being on a COCOM's Integrated Priority List was discussed. Another comment was that something like a Joint Urgent Operational Need and a rapid fielding mechanism might be needed to overcome obstacles in the usual acquisition process. Mr. Jaggers discussed institutional processes that need to be considered in order to get a MAJCOM to "make a bet" on transitioning DEW technology. Several participants commented on their views of challenges to DEW transition because of these institutional processes (see Theme 1.5 in Chapter 2 for illustrative examples). There was extended discussion of difficulties in getting from DEW demonstration projects to prototypes that might interest the operational Air Force enough to win investments in operational capability. Which DEW applications might be ripe for moving to a prototype, in addition to CHAMP, was another topic of discussion (see Theme 3.1).

Results from a Directed Energy Task Force (Directed Energy Senior Advisor Group)
Maj Gen David J. Scott (USAF, Ret.), Workshop Planning Committee Member

Maj Gen Scott described the role of the Directed Energy Senior Advisor Group (DESAG) and its latest meeting on April 11, 2012. The value of a group like the DESAG, he said, is that it helps to address the problem of maintaining continuity of key players as leaders move to new postings. At the April 2012 meeting, the DESAG discussed the merit of conducting an AoA for a counter-electronic warfare application of HPM. In response to a question, he said there has not been an analysis of alternatives done for any HEL application.

In Gen Scott's opinion, DEW is not currently "on the radar screen" for senior Air Force leadership. There are DEW capabilities already out there for certain applications, he said, but what would be something useful that can be suggested? If the right DEW capability can be identified and be supported by knowledgeable advocates, he continued, it will still need to go through the Air Force Requirements Oversight Council (AFROC) or the Joint Requirements Oversight Council (JROC) to move forward with an Advanced Concept Technology Demonstration (ACTD) or JCTD, respectively. Dr. Fender commented (by telephone) that Gen Hostage, the ACC Commander, is interested in what the workshop's participants have to say about what directions make sense for DEW. Aircraft self-defense is no longer necessarily his top priority, she added.

U.S. Pacific Command (PACOM)
Mr. Ken Bruner, S&T Advisor, PACOM

Mr. Bruner commented on the CHAMP Joint Concept Technology Demonstration (JCTD), for which PACOM was the proponent, and on how the demonstration is viewed from a COCOM perspective. To verify the simulation models, two flight demonstrations were initially planned, but only one flight demonstration was actually flown. The results were even better than indicated by the earlier ground testing of the system. The missile flew as planned and was successfully guided, HPM was successfully pointed by rolling the missile, and the payload worked. Most of the effects on the targeted electronic systems were temporary, Mr. Bruner said, but the cyber realm liked that because it forced shut down and restart of the systems at a given time. From PACOM's perspective, he continued, the system would be useful in various Pacific Air-Sea Battle scenarios.

PACOM is also interested in testing CHAMP against big-ship electronics and favors going forward with the proposed rapid acquisition program for 32 missiles, plus full life-cycle cost funding and contractor logistical support for 10 years. In general, PACOM is supportive of where ACC is going with CHAMP and other HPM capabilities, including in-band options. When asked about PACOM interest in HEL capabilities, Mr. Bruner mentioned LaserComm, a pulsed illuminator for missile defense, and long-range IRST as HEL concepts of interest. IRST, he said, could give both a combat ID capability (including a solution for long-range counter-DRFM) and in-band sensor blinding, particularly to defeat air-to-air missiles.

Navy DEW Initiatives
Mr. Peter Morrison, Research Program Office, Office of Naval Research (ONR)

ONR handles 6.1 to 6.3 programs, Mr. Morrison said, with some 6.4 that is directed to transitions to acquisition programs. ONR's DEW focus for the near term is on solid-state lasers. The Navy's work on the Free Energy Laser (FEL), which is a longer-term prospect, is in collaboration with the Department of Energy national laboratories. The Naval Expeditionary Combat Command (NECC) has a new focus on DEWs for Navy applications, he said. Operators from NECC participate on the Navy's Integrated Product Teams (IPTs), and review and acceptance by an IPT is important to give feedback to the S&T community or to move forward into transition and acquisition.

DEW is seen as important as a counter to RPAs and small boats, Mr. Morrison continued. Because DEW is seen as having a major role in providing "magazine depth" for certain threats, including RPA and small boat swarms, it is viewed as putting Navy defensive systems against these threats on the good side of the capability versus cost curve. The Navy sees HEL as a nearer-term solution than high-power radio frequency radiation to counter the small boat/RPA threat. He listed HEL benefits to naval warfare and discussed ONR's Laser Technology Roadmap. ONR sees the Navy's solid state lasers as being at Technology Readiness Level (TRL) 5; they need to get to TRL 6 for transition, he said. FEL, which ONR views as a long-term possibility, gives capability for agile, narrowband capability in which both atmospheric particulate and moisture absorption of the laser beam can be minimized. FEL would have to be on an aircraft carrier-size hull. Mr. Morrison agreed that DEW R&D is being pushed toward delivering prototypes and not just concept/technology demonstrations. Feasibility studies indicate that a DDG-51 hull would have space to mount a 100 kW solid-state HEL system. Another concept that ONR is exploring is a high-energy fiber laser (HEFL) system mounted in a marine helicopter for applications such as mine-clearing. For use by the Marine Corps, ONR is looking at ground based air defense with a mobile HEL system, and Mr. Morrison discussed the trade-offs between HEL and kinetic weapons for air defense. The current CONOPS for a ground-based air defense HEL is "shoot and scoot," where the vehicle-mounted system stops to shoot, then moves. He also discussed ONR interests in counter-DEW warfare.

Discussion of Day 1 Presentations
Gen Ronald E. Keys (USAF, Ret.), Session Chair

The participants began the Day 1 discussion session with comments comparing the directions and relative progress of Navy work on DEW with Air Force directions and work status. One representative comment was that, overall, the operational Navy seems to have accepted DEW applications more than the operational Air Force has. Several participants observed that the Navy appears further along in fieldable development for its focal DEW applications, such as responding to small-boat swarms, air/land-to-sea missile swarms, or mine clearing, which do not have the SWaP constraints that apply to potentially important Air Force roles for DEW. With respect to differences between the Navy and Air Force in general programmatic and technology development approaches, several participants noted a greater emphasis in the Navy's programs on getting to prototypes. One participant noted that the

SWaP constraints are very important. The SWaP constraints for the Army and Navy applications can be met with current technology, whereas the Air Force constraints [essentially, feasibility for an airborne weapons subsystem] are not met with current technology.

Another participant said that, other than HPM applications [such as the CHAMP counter-electronics role], he did not see a clear focus yet for Air Force DEW applications. As an example, he questioned whether the Air Force really needs to have a HEL system on large aircraft for self-protection: "What is the requirement?" he asked. However, another participant countered that HEL applications such as advanced Directed Infrared Countermeasures (DIRCM) and counter Digital Radio Frequency Memory (counter-DRFM) were reasonable near-term airborne applications that could be important to the Air Force.

Many of the participants in this discussion made comments in favor of the importance of a ground-based HEL system for air base defense as a near-term DEW application. However, as several of them added, base defense is a mission assigned to the Army and is not an Air Force mission. (See Theme 2.) Gen Keys summarized his perception of the extended discussion on this topic as follows, "Recognition of assigned roles and missions in a joint context are important; however, the Air Force has a vital interest in air base defense and should therefore provide support to ensure that the DE base defense capability on which the Army is working will provide operational systems that cover Air Force needs." There was a range of views on how the Air Force could best pursue this "vital interest." During the third-session review of this discussion, a participant disagreed with the implication that the Air Force should "provide support" [i.e., provide funding] and noted that doing so would dilute Air Force resources on Army missions. A specific suggestion from one participant during the second-session discussion was that the Air Force might ensure that the Army's target set for base defense includes RPAs, particularly in swarms that could present a threat to aircraft parked at an airfield. There was discussion (plus additional comments from Session 3 participants) of whether the RPA swarm threat or other air attacks (cruise missiles and intermediate range ballistic missiles were specifically mentioned) might be something that the Air Force would need to address itself, if Army-supplied systems were not demonstrably effective against such threats. Several participants expressed interest in hearing more about the Army's HEL program for defense against airborne threats to ground bases (the Army C-RAM mission) and more about the DARPA HEL program. Presentations on both of these programs were included in the third session.

The chair wrapped up the discussion during this session with the comment, "It seems [to me] that the Navy sees clear and present threats that DEW can help defeat. For the Air Force, other than CHAMP [the nonkinetic, counter-electronics mission], there is not a clear requirement or threat that is best defeated by directed energy." To the point made by another participant that perhaps there was an operational need for directed energy to defeat RPAs and other airborne assets, the chair replied that, at least for a DEW system carried on a fighter aircraft, directed energy does not yet provide a deep magazine capability. "A comparison of alternatives for a HEL weapon system [as one of the alternatives] on a fighter has not been done," he added, "so [HEL] seems to be a solution looking for a problem." This view from the chair of the direction of the discussion was considered further in the opening and wrap-up discussions on the second day. During the third-session review of discussion highlights from prior sessions, several participants expressed agreement with the chair's summary of the discussion or with specific portions of it. But one third-session participant noted that the DARPA presentation on the ENDURANCE program [given during the third session] was a good example of a pod-based HEL system that could be carried on a Reaper RPA or other aircraft. Another counter-comment was that, during a simulation-based analysis of alternative self-protect systems for the F-15, a HEL system was assessed to be more effective than a gun system.

Tuesday, March 19, 2013

Session 2, Day 2: Welcome and Recap of Day 1

Gen Ronald E. Keys (USAF, Ret.), Session Chair

Gen Keys welcomed the participants to the second day of Session 2 of the Workshop on Assessment of Directed Energy Research and Development for U.S. Air Force Applications. He noted the presentations that had been given on Day 1 and summarized highlights of the participants' discussions. Returning to the issue raised in the Day 1 wrap-up discussion of what might be the key DEW applications—and particularly HEL weapons—for nearer-term Air Force applications, a participant noted that aircraft-carried HEL systems for antisensor and combat ID applications had not been discussed yet in Sessions 1 and 2. The chair agreed with the point and added that the participants also needed to think about counter surface-to-air-missile (counter-SAM) applications, applications to counter surface-to-air aiming subsystems and counter-optics, and counter-sensor applications generally. For those applications, he explained, the energy and power for effectiveness does not need to be as high as for vehicle-skin defeat mechanisms. A HEL in the 25-30 kW range might suffice, he said, but adequate stand-off range would be a critical requirement.

In the third-session review of Session 2 highlights, several participants agreed with the chair's comment that the counter-sensor and combat ID applications for HEL are good missions for the Air Force. One participant noted that the range for such "lower-power" HEL systems might be even lower, perhaps down to 10 kW. Another comment on the last statement was that weather (e.g., clouds) remains an issue [in addition to stand-off range], and a third participant suggested that critical requirements would be "adequate range, tracking, and pointing."

Boeing DEW Initiatives: CHAMP JCTD Overview and Follow-on Next Steps

Mr. Keith Coleman, CHAMP Program Manager, Phantom Works, Boeing

Mr. Coleman prepared an abstract of his presentation for inclusion in this report; it is incorporated below as Box 3-6.

BOX 3-6

CHAMP JCTD Overview and Follow-on Next Steps
Mr. Keith Coleman, CHAMP Program Manager, Phantom Works, Boeing

The CHAMP (Counter-electronics HPM Advanced Missile Project) program designed, developed, and flight-tested a missile-deployed HPM weapon system. CHAMP allows warfighters to target the electronic systems of an adversary and disable their ability to function. By using an HPM warhead, CHAMP provides a highly effective but nonlethal counter-electronics capability to minimize or eliminate target collateral damage. The CHAMP team integrated a spectrum of complex systems (flight controls, power distribution, HPM source, targeting systems, etc.) creating solutions in a number of flight control and materials fields on their way to a successful flight test on October 16, 2012, at an Air Force flight test range. Problems that were overcome included pulse power generation, voltage isolation, electromagnetic protection of vulnerable missile flight systems (prevention of self-kill), precise targeting, weapon pointing and triggering techniques, microwave beam propagation to target, prediction and verification of microwave weapons effect on targeted electronics, and development of the concept of operations and operational analysis of a new class of missile.

The CHAMP team, which included members from the Air Force, Boeing, Sandia National Laboratories, and Raytheon, developed this engagement capability with the potential to minimize reconstruction requirements post-conflict. The weapons system provides the capability for multiple, nonlethal, pinpoint airborne strikes against targets that were previously denied due to proximity to, or co-location with, noncombatants. A rapid acquisition program was proposed and is available to provide assets for warfighter use in the near term.

Raytheon DEW Initiatives: CHAMP Payload and Effects Testing
Dr. Peter Duselis, Raytheon Company

Dr. Duselis began by noting that the CHAMP payload in the 2012 flight test was at TRL 7. The current HPM source used in CHAMP, ACES, requires a vacuum pump to maintain the system, and the vacuum condition in the system is the critical factor for system shelf life. It will hold a vacuum for a year, then requires 24 hours to pump down to the working pressure. Dr. Duselis also described the software modification that would allow CHAMP to used the onboard power of the B-52 carrying it before release. He reviewed the flight test results, which Mr. Bruner and Mr. Coleman had also covered, but provided more detail on the level of effects on the target electronic systems. Dr. Duselis then reviewed the major components of the proposed CHAMP payload improvements and explained their value. Overall, the proposed upgrades would provide higher effective radiated power, eliminate or simplify subsystems, and multiple pulses on target. For the proposed two-pulse option, pulses could be as close together as 10 ms.

HPM effects testing is necessary to answer questions about HPM effectiveness against potential targets. In June 2011, Raytheon and AFRL entered into a cooperative research and development agreement (CRADA) to enhance the existing database for HPM effects. Dr. Duselis presented details of the effects testing done at White Sands Missile Range, including the effects of the ACES-derived HPM beam on various targets of interest for Air Force counter-electronics attack. The workshop participants and Dr. Duselis discussed potential uses and CONOPS for a CHAMP-like counter-electronics capability.

General Atomics DEW Initiatives
Dr. Chris Pehrson, General Atomics

Dr. Pehrson presented the briefing for Dr. Michael Perry, Vice President, Laser and Electro-optics, General Atomics, who had a last-minute schedule conflict. The presentation began with an overview of the emergence of electrical power systems suitable for DEW systems on jet-powered armed reconnaissance RPAs such as Predator and Avenger. The second topic was the integration of optical radar (LIDAR) into widely deployed EO/IR turrets. The third topic was on the emergence of tactical laser weapons in the DARPA HELLADS program, including the second-generation HEL Avenger system and the third-generation Laser Weapon Systems Module (LWSM).

Dr. Pehrson described the evolution of Predator/Avenger RPAs and characteristics of the latest generation. He then presented characteristics of the Spi-3D LIDAR and its use of three-dimensional (3D) voxels to get precise ranging and targeting data over the entire field of view of the LIDAR, enabling real-time precsion-targeting of multiple targets. This capability can be installed in existing Predator airframes with only minor modification to the sensor ball. On electric lasers for tactical laser weapon applications, Dr. Pehrson noted both advantages (ultra-precise effects, deep magazine, low cost per shot) and constraints (aimpoint must be maintained on moving target for 3-4 seconds, not an all-weather capability) of tactical laser weapons. He contrasted the SWaP requirements of the new distributed-gain electric lasers with the SWaP requirements of the Airborne Tactical Laser. This new electrical laser would potentially fit entirely within the internal bay of an Avenger RPA and has an output of 100 kW. The recharge time is 2-4 minutes versus more than an hour for the ATL, and the magazine depth is unlimited.

Next, Dr. Pehrson discussed the DARPA High Energy Liquid Laser Area Defense System (HELLADS), which aims to take a distributed gain solid state tactical laser system from concept to weapon system demonstration with a 150 kW output power objective. A HELLADS Demonstrator Laser Weapon System (DLWS) with 150 kW output is now complete and is scheduled to be delivered to White Sands Missile Range on June 30, 2013.

RELI is a second-generation HELLADS that is now undergoing laser performance validation. First light for the laser head, Dr. Pehrson reported, was on March 19, 2013. Its size and weight are small enough to enable it to be carried under a Predator C Avenger RPA (the HEL Avenger concept), and its output power is about 100 kW. Unit cost of the HEL Avenger laser system will be about $20 million to $22 million. Dr. Pehrson presented information on modeling and analysis tools that will be used to model efficacy of the RELI system, when mounted on an airborne platform, against a range of potential targets. The next generation beyond the RELI unit cell in the HEL Avenger system will be a 150 kW HELLADS DLWS laser, incorporated in the Airborne LWSM. The long-term objective for Airborne LWSM is to successfully attack and take out an integrated air defense system.

Navy Initiatives (China Lake)

Mr. Tom Glover, Head, Emergent Weapon Systems, Directed Energy Weapons, Naval Air Systems Command (NAVAIR)

Mr. Glover gave his briefing by telephone, with briefing slides displayed at the workshop meeting room. He reviewed the NAVAIR roadmap for HEL weapons R&D and transition. The roadmap projects reductions over time in the SWaP for different tactical platforms. A key issue for reducing SWaP is the ruggedization needed so that systems can withstand carrier-based take-off and landing. Multiple departments within NAVAIR are involved in work on integrated laser systems for different Navy platforms. Mr. Glover also reviewed the NAVAIR roadmap for HPM weapons R&D. For the Navy, HPM is a niche application, compared with HEL weapon systems, which are seen as being used on many platforms for different applications.

Turning to countermeasures development, Mr. Glover discussed anticipated near-, mid-, and far-term HEL and HPM weapon systems that potential adversaries may have. The NAVAIR countermeasures group has developed a "laser technology landscape," which Mr. Glover presented. He discussed laser threats to Navy sensors and NAVAIR's plans to work with AFRL on sensor protection. He also discussed hardening required to defend electronics against an HPM threat environment, including NAVAIR plans for live fire test and evaluation for survivability against HPM attack. He closed the presentation with a summary of joint Air Force/NAVAIR activities.

Technological Assessment of Foreign Directed Energy

Mr. Michael H. Bernard, Senior Intelligence Analyst, AFRL

This assessment covered both radio frequency weapons (RFW) such as HPM and HEL weapons. Mr. Bernard described what is known about the technical status of work in specific foreign R&D institutes on both RFW and HEL weapons.

Discussion of Day 2 Presentations and Plan for Session 3

Gen Ronald E. Keys (USAF, Ret.), Session Chair

The following high-level points were suggested by one or more of the Session 3 participants who were present for this final discussion. Elements of these points have been incorporated in the thematic overview in Chapter 2.

- Many of these participants who spoke agreed that some DEW technologies are ready for tactical application, in the sense that some capability does work and could have operational value. The Army and Navy have demonstrated prototypes. What is missing is what some described as a capabilities-based assessment and others described as "an operational need backed up by an Analysis of Alternatives." They agreed that the focus of this assessment, however structured, should be to determine how DEW capability could complement, expand, or replace kinetic weapon alternatives.

— Several emphasized that this assessment, however done, should focus on what can be done now and what can move now from demonstration to prototypes. (See Theme 3.1 for more on this point.)

— Many participants thought the compelling missions for DEW need to be identified. One suggested that an important area to consider is the potential integration of kinetic weapons and DEWs—for example, in a campaign-level analysis of how they would work together. A specific example suggested by a participant was target marking, with a DEW to prepare and mark a target for a follow-up kinetic attack (this was not intended to mean just a laser-designation application). Another participant mentioned counter-sensor and combat ID roles as mission area where DEW and kinetic weapons could work together.

- With respect to HPM weapons, many of these participants thought it important to go forward with CHAMP acquisition.

  — For example, one participant suggested taking advantage of the interest in CHAMP from at least one COCOM—get an initial capability out there and then look at how to improve it.

  — Several participants who favored going forward on CHAMP did not favor doing all of the proposed upgrades as part of an initial buy. One of them explained that the upgrades, although valuable, introduced technical and programmatic/schedule risks that were best avoided at this point. A block-buy evolution of capability was suggested by one participant.

- There was considerable discussion, with a range of views expressed, about the near-term prospects for laser DEW capabilities. The views expressed on this topic are represented in the Chapter 2 discussion of Themes 1.2, 2, 3.1, 3.2, 3.3, and 3.4.

- With respect to Blue Force defense against Red Force DEW systems, a number of these participants suggested that an assessment (again, a capabilities-based assessment or something like an Analysis of Alternatives) was needed of counter-DEW options. What is (are) the most effective way(s) to counter Red Force DE capability, either offensive or defensive?

  — An important issue for counter-DEW, several agreed, is the relative cost of alternatives, including kinetic as well as nonkinetic (DEW) alternatives. As one participant expressed this issue, "What other capability would be worth giving up to get a DEW capability for counter-DEW?"

  — In the context of Blue Force airborne DEW weapons to counter Red Force defensive, ground-based DEW weapons, several comments suggested that an open question is whether airborne DEW weapons would be successful in "jousting" with the larger, more powerful Red Force systems. Other comments suggested that countering Red Force defensive DEW systems might be best done by CONOPS that combined DEW and kinetic weapon capabilities.

SESSION 3
APRIL 24-25, 2013
WASHINGTON D.C.

Wednesday, April 24, 2013

For this session, the workshop planning committee developed guidance specific to individual invited presenters on questions to be answered or topics to be addressed. The guidance given for the presentations from the Air Force Air-Sea Battle Office, the Army, and DARPA was specific to each of those topics and is included in the synopses below. The three presenters from industry (Lockheed Martin, Northrop Grumman, and Boeing) were given the same guidance:

> Describe fairly near-term (5 years or less) laser DEW concepts that address Air Force missions. Include effectiveness for Air Force missions and CONOPS thoughts. To this end, the workshop participants do not need details about the technology, per se. These missions could involve low-to-medium power lasers (e.g., 10 or 10s of kilowatt power levels). NOTE: HEL-JTO, MDA, AFRL, MIT/LL, DARPA HELLADS, Army SMDC, Navy ONR, ACC, and General Atomics have already provided briefings. The workshop participants would prefer that the briefing not be proprietary since a summary of the briefing will go into a National Academy document. The presentation can be up to the Secret level.

Welcome and Introductions
Gen Ronald E. Keys (USAF, Ret.), Session Chair

Gen Keys opened the meeting and asked the participants to introduce themselves. He then proceeded directly to the first topic, which was to discuss and comment on the rapporteur's draft themes and major points from the first two sessions.

Discussion of Themes and Points from Sessions 1 and 2
Dr. Robert Katt, Rapporteur

As he did at the beginning of Session 2, Dr. Katt reviewed with the Session 3 participants a draft of highlights (key themes and points made by individual participants) from the previous two sessions. The participants' comments were captured for incorporation into the final version of themes presented in Chapter 2 of this workshop report.

Air-Sea Battle Concept of Operations
Col Jordan Thomas, HAF/A5X

The specific guidance sent to Col Thomas was to address the following questions:

1. Which service has the lead for protecting such fixed bases as Guam?
2. Has this [protection of fixed bases] been coordinated with the Army for their base protection [mission]?

3. Are directed energy weapons (DEW) being considered in the trade space for that protection?
4. What Air-Sea Battle operational needs could potentially be met by DEW?
5. Have any threats been identified by [the Air-Sea Battle organization] that suggest potential Blue Force DEW applications?

Col Thomas noted that Air-Sea Battle is not new for either the Air Force or Joint forces, but it has not been updated since about 2000. With respect to question 1, he said that, according to JP-3-10 doctrine, the lead service for protecting a fixed base depends on the command structure set up by the responsible COCOM. There is no single executive agent for area air defense coordination. He then described some of the current planning for defense of Guam. On question 2, Col Thomas said that the Army and Navy have their own DEW programs. There is no coordination between them and the Air Force.

With respect to question 3, Col Thomas said that DEW is not included in the defensive solutions being considered by the Air-Sea Battle Office. He agreed that it could be a capability of interest. Base defense is the number-2 priority, and the Air-Sea Battle Office sees a need for a capabilities-based assessment for base defense. But to his knowledge, neither the Air Force nor the Joint Staff have done or considered such an assessment that includes DEW. In response to a later question, he said that DEW capabilities are not included in Air Force and Joint wargaming. In response to question 4, Col Thomas said that the principal long-term threat for his office is A2AD (anti-access/area denial) by an adversary. The trend toward higher precision in Red Force offensive systems means that A2AD is becoming a concern with respect to an increasing array of potential adversaries. In response to a question, he said that nearer-term threats, such as having to operate in Southwest Asia or South Asia, are also a concern. Col Thomas also highlighted the key principles that the Air-Sea Battle office is applying to planning for various scenarios. He responded to participants' questions about how specific DEW capabilities might be useful.

Results of Army DEW Analysis of Alternatives (AoA)

Dr. Kip Kendrick, Chief, Directed Energy Division, U.S. Army Space and Missile Defense Command, Army Forces Strategic Command

For this briefing, Dr. Kendrick was sent the following guidance from the workshop planning committee:

1. Describe the AoA that was done for the counter rocket, artillery, and mortar base defense mission and the outcome of the AoA.
2. Why was the final decision for the go-ahead not the alternative that came out on top for the AoA?
3. Summarize what the AoA found were the advantages of DEW over alternative kinetic solutions.

Dr. Kendrick started by providing more background on the IFPC II AoA than he had given in his presentation to Session 1, including the motivation and the specific scenario and parameters that had to be addressed. One requirement was to provide complete life-cycle cost estimates for two battalions of each alternative capability. The DEW options did well on cost because they did not have to replace bullets or missiles. With respect to the ratio of threat kills to the total number of threat targets presented, the 100 kW HEL system had the highest score, followed by the EAPS missile option and the 50 kW HEL option. For the HEL systems, the beam control system and laser head were identified as having moderate technical risk, given their developmental status. Also, the DEW options were viewed as unsuccessful in fog, rain, and sandstorms.

Since the AoA, the Army has expressed increased interest in cruise missile defense as part of defending a FOB. To Dr. Kendrick's knowledge, HEL systems were not included in the Army's analysis of cruise missile defense alternatives, although he offered to provide data to them. In response to participants' questions about his sense of the Army leadership's resistance to HEL for base defense, Dr. Kendrick recounted the story of how the Army refused to consider main battle tanks in World War I. The Army was not interested until after Winston Churchill set up the Land Ship Commission and successfully deployed British tanks in trench warfare. He sees DEW systems as in an analogous position, with no specific formal requirement and no acquisition program. Dr. Kendrick is working on requirements for a couple of DEW platoons for the Army. To get to an advanced prototype, he needs an advanced beam control system. He is looking for funding to do a full-up prototype system for a Joint prototype demonstration with Air Force and Navy support, but not a JCTD.

DARPA Airborne DEW Initiatives

Dr. Joseph Mangano, Program Manager, Advanced Lithograph Program, Microsystems Technology Office, DARPA

The specific guidance sent to Dr. Mangano for this briefing was as follows:

1. Describe your laser program, including its progress to date and its future plans, as well as its relevance to the Air Force.
2. Provide examples of Air Force operational needs that could possibly be met by low- and/or medium-power lasers.

Dr. Mangano's presentation focused on the DARPA Endurance and Flash programs, both of which use fiber laser arrays. The technological objective is to achieve "ultra-low SWaP" at 2 kg/kW specific power and a size of 1 m^3 per 100 kW. The SWaP is critical for both these programs, he said, and the integrated system must be conformal to aircraft platform constraints. As applications of interest, he mentioned aircraft self-defense and mid-course ballistic missile defense.

In both technologies, the light reflected from the target (the return) is used to tune the combined multiple beams from 3-kW laser fibers. The individual 3 kW fibers operate at 35 percent wall-plug efficiency. This works well enough that the beam at the target is close to the diffraction-limited spot size. Although Dr. Mangano said the limits of scalability for this beam-combining approach are not clear, he can do a hundred or more fibers together and believes the technology is scalable to a 1 MW output beam.

The Endurance program is aimed at defeating EO/IR guided surface-to-air missiles that threaten airborne platforms at altitudes up to 50,000 feet. Dr. Mangano described what he sees as the capabilities of current state-of-practice Red Force surface-to-air missiles, as well as some of the emerging threat capabilities, and he discussed how an Endurance system would likely perform against these threats. The program is planning to use a HEL pod that is small and light enough to fly on legacy aircraft as well as advanced next-generation aircraft.

The Flash program is focused on defeating existing EO/IR guided air-to-air missiles, as well as surface-to-air missiles, with a beam power of up to 300 kW. The Excalibur program was the predecessor of Flash. Dr. Mangano described the key technologies and the program roadmap to go from the current Excalibur program to Flash capability that includes a 30-shot magazine. In closing, Dr. Mangano gave the current milestones for the Endurance and Flash programs between FY 2014 and FY 2017.

Lockheed Martin HEL Efforts

Mr. Thomas Burris, Lead Systems Engineer, Directed Energy Program, Lockheed Martin

Mr. Burris prepared an abstract of his presentation for inclusion in this report; it is incorporated below as Box 3-7.

Northrop Grumman Laser Efforts

Mr. Kenton Ho, Director, Laser Products Directed Energy, Northrop Grumman

Mr. Ho began with an overview of Northrop Grumman's current (2013) HEL programs. Northrop Grumman is now working with solid-state lasers in the 100 kW class. He described the

BOX 3-7

Lockheed Martin HEL Efforts

Mr. Thomas Burris, Lead Systems Engineer, Directed Energy Program, Lockheed Martin

Laser weapons can create a wide range of effects that span the find, fix, track, target, engage, assess (F2T2EA) targeting chain. There are advantages in time (proactive effects, counter salvos), favorable cost-exchange ratios, precision, reduced collateral damage potential, and magazine depth (e.g., increase mission duration). Lockheed Martin has performed numerous airborne operations analyses and aircraft integration studies that have led to the following conclusions: (1) directed-energy weapons will not replace kinetic capabilities, but rather directed-energy and kinetic weapons are complementary and synergistic; (2) high-energy laser weapons can help fill numerous capability gaps spanning the breadth of offensive and defensive missions; and (3) multiple, compelling applications have emerged as opposed to a single "killer application." A final point worth noting is that a first-generation laser weapon needs to provide sufficient output power and beam quality to enable operational capabilities that are commensurate with the high cost of the laser weapon system and aircraft integration.

Maritime Laser Demonstration, which involved engaging small-boat threat scenarios out to 6,000 yards. He also described the attack performance of a 10 kW HEL pod mounted on a RPA.

The presentation included an overview of Northrop Grumman studies of HEL weapon systems on tactical aircraft for defense against surface-to-air missiles in an A2AD scenario. HEL systems from 25 kW to 150 kW were considered and were evaluated against low-observability alternatives. Mr. Ho emphasized the increase in aircraft "autonomy" and flexibility provided by a multiuse tactical HEL weapon system. In closing Mr. Ho said that ACC needs to give industry a clear requirement, then let industry work out the best, most affordable way to meet it.

Boeing Airborne DEW Systems
Dr. Harold Schall, Boeing; Mr. Michael Black, Boeing

Dr. Schall prepared an abstract of his presentation for inclusion in this report; it is incorporated below as Box 3-8.

General Discussion on Day 1 Presentations
Gen Ronald E. Keys (USAF, Ret.), Session Chair

Gen Keys asked the participants for comments on the Day 1 presentations. The participants' comments have been selectively incorporated in the rapporteur's overview of workshop themes (see Chapter 2).

BOX 3-8

Boeing Airborne DEW Systems
Dr. Harold Schall, Boeing; Mr. Michael Black, Boeing

Boeing has made significant progress over the past decade maturing HEL system technologies and demonstrating the capabilities of integrated HEL systems. Demonstrations such as the Airborne Laser's lethal shootdown of a boosting missile and the Airborne Tactical Laser's precision strike engagements against static and dynamic targets have clearly shown the unique capabilities offered by a directed energy weapon. Based on the current state of HEL system technologies, Boeing has identified three Air Force mission areas that it believes can be addressed in the fairly near term by low- to medium-power lasers. These mission are (1) Aircraft Self Protection from Advanced EO Threats, (2) Precision Air-to-Ground and Air-to-Air Engagements and (3) Long Range Detection and Engagement from Airborne Platforms. The mission space addressable in the near term spans a wide space and multiple platforms. This presentation describes the directed-energy concepts and capabilities that address these Air Force mission areas.

BOX 3-9

CHAMP Advocacy Brief

Dr. David Robie, Office of the Chief Scientist, ACC

The Counter Electronics High Power Microwave Advanced Mission Program (CHAMP) Joint Capabilities Technical Demonstration just completed a 3-year, $70 million development effort with outstanding results. The Air Force would like to leverage this positive outcome and investment to provide the warfighter with game-changing capabilities. This presentation discusses the requirements for the capability, the results of the operational utility assessment, a discussion of the survivability of the weapon and an evaluation of countermeasures to the weapon's effects. It concludes with a presentation of courses of action to field this asymmetric and disruptive capability.

A Retrospective Look at the Northrop Grumman Laser Weapon System Initiative (LWSI)

Dr. Thomas Romesser, Independent Consultant, Member of the Workshop Planning Committee

To provide the workshop with a historical perspective on the degree to which technology HEL status forecasts and envisioned DEW applications have (or have not) changed in the past decade, Dr. Romesser presented and discussed a PowerPoint briefing he had prepared on the Northrop Grumman Laser Weapon System Initiative (LWSI) in 2003.

ACC Draft Briefing on CHAMP

Dr. David Robie, Office of the Chief Scientist, ACC

Dr. Robie prepared an abstract of his presentation for inclusion in this report; it is incorporated below as Box 3-9.

Discussion of Potential Workshop Themes and Key Points

Gen Ronald E. Keys (USAF, Ret.), Session Chair

Gen Keys asked the participants for their general reflections on this session and the workshop as a whole (for those who had attended other sessions). He also asked for any additional comments they had on the draft themes and major points that had been presented and discussed on Day 1 of this session. The participants' comments have been selectively incorporated in the rapporteur's overview of workshop themes (see Chapter 2).

Appendix A

Biographical Sketches of Committee Members

Ronald E. Keys (Gen, USAF, Ret.), *Chair*, is president of RK Solution Enterprises, LLC. Gen Keys retired from the Air Force in November 2007 after completing a career of more than 40 years and command of eight different organizations. His last assignment was as commander, Air Combat Command (ACC), at the time the Air Force's largest major command, consisting of more than 1,200 aircraft, 27 wings, 17 bases, and 200 operating locations worldwide with 105,000 personnel. Under his leadership, ACC organized and stood up the Air Force's first RPA Wing and first Network Warfare Wing. Gen Keys holds a bachelor of science from Kansas State University and a master's degree in business administration from Golden Gate University. In addition to numerous professional military education courses, Gen Keys has participated in the National and International Security Seminar, JFK School of Government, Harvard, as well as Leadership at the Peak, Center for Creative Leadership, in Colorado Springs. In September 2007, he received the Air Force Association's most prestigious annual award, the H. H. Arnold Award, as the military member who had made the most significant contribution to national defense. Upon retirement he was also selected as the Air Force Reserve Officer Corps' first AFROTC Distinguished Alumni award recipient. Gen Keys is a command pilot with more than 4,000 flying hours in fighter aircraft, including more than 300 hours of combat time in Southeast Asia.

William L. Baker is an independent consultant since 2010 when he retired as chief scientist for the Air Force Research Laboratory's directed energy directorate. Prior to this position, Dr. Baker was senior scientist for high-power microwaves at the Air Force Phillips Laboratory. Concurrent with this position, he was a special technical consultant on the staff of the Joint Chiefs of Staff. Dr. Baker is an expert in directed energy, plasmas and pulsed power, nuclear science, space situational awareness and protection, weapons, technical intelligence, and directed energy effects on humans. He is a fellow of the Directed Energy Professional Society (DEPS) (2010) and a fellow of IEEE (1999) and is a recipient of the following awards: Air Force Outstanding Civilian Career Award (2009), Distinguished Senior Professional Presidential Rank Award (2006), Meritorious Executive Presidential Rank Award (2003), and Department of Defense (DoD) Distinguished Civilian Service Award (1999), among others. He holds a PhD in nuclear physics from Ohio State University.

Jacqueline G. Gish, a member of the National Academy of Engineering, is an independent consultant. Dr. Gish retired from Northrop Grumman Aerospace Systems in 2011, where she had been a technical fellow in technology development. Prior to that, Dr. Gish served as the director of directed energy technology for Northrop Grumman Space Technology. Dr. Gish also served as a department manager, program manager, and business manager in the laser area. She was elected as a fellow of DEPS in 2004 and served on the National Research Council's (NRC) Air Force Studies Board (2004-2010) and the Advisory Board of Scientific Advisors for DEPS (2005-2007). Dr. Gish has authored or co-authored more than 30 publications and presentations and holds three patents. Dr. Gish holds a PhD in chemical physics from the California Institute of Technology and an MBA from the executive program at the University of Southern California.

Matt L. Mleziva is currently the president of Wildwood Strategic Concepts, LLC, a strategic management company in Westford, Massachusetts. Mr. Mleziva has led Joint teams for the Office of the Secretary of Defense that developed recommendations projected to save millions of dollars annually. He guided U.S. Air Force Networked Tactical Communications efforts into a single Joint program with the U.S. Navy. Mr. Mleziva has a proven track record of achieving cost, schedule, and performance goals across organizations covering a wide range of information system technologies for a diverse customer base. He acquired space, air, and electronic systems for DoD, the U.S. government, and foreign nations. Mr. Mleziva has a demonstrated capability to utilize emerging information technology and promote commonality and interoperability in combat systems. He developed an ultra-streamlined acquisition strategy in response to urgent Air Force operational needs. Mr. Mleziva is the recipient of several awards, including the Presidential Meritorious Executive Rank Award and the Air Force Outstanding Civilian Career Service Award. He holds a post master's degree in electrical engineering and an M.S. in electrical engineering from the Massachusetts Institute of Technology. Mr. Mleziva is a member of the NRC's Air Force Studies Board and a past member of the Committee on Examination of the Air Force Intelligence, Surveillance, and Reconnaissance (ISR) Capability Planning and Analysis (CP&A) Process and the Committee on Evaluation of U.S. Air Force Preacquisition Technology Development.

Thomas E. Romesser, a member of the National Academy of Engineering, is an independent consultant. Until the start of 2012, Dr. Romesser was chief technology officer for Northrop Grumman Aerospace Systems and sector vice president of Aerospace Systems, a premier provider of manned and unmanned aircraft, space systems, missile systems, and advanced technologies critical to our nation's security. In this role he provided senior leadership representation with customers, universities, industry, and the rest of the corporation. He also was responsible for technology development to support future programs while maintaining close linkage to legacy programs. Prior to this assignment, Dr. Romesser was sector vice president and general manager of Technology and Emerging Systems Division for Northrop Grumman's former Space Technology sector. In this role, he was responsible for the development and execution of Space Technology's strategy to support both near- and long-term business objectives, system enhancements, and technology leverage for new business pursuits. He oversaw activities of the Directed Energy Systems and Advanced Concepts organizations as well as the Space Technology Research Laboratories. Previously, Dr. Romesser was vice president of Technology Development, responsible for the identification, development and acquisition of Space Technology's strategic technologies, and he managed discretionary investments in technology and product development. Dr. Romesser joined Northrop Grumman via the acquisition of TRW in 2002. A vice president since 1998, he previously served as vice president and deputy of the Space and Electronics Engineering organization. Prior to that, he was vice president and general manager of TRW's Space and Technology Division, responsible for spacecraft hardware and software engineering, manufacturing, test, and space vehicle production, as well as chemical and solid-state laser design and development, sensor systems, space and tactical propulsion systems, and research in the physical, chemical, and engineering sciences. Since joining the company in 1975, he has been involved in the development and management of a broad range of high-technology capabilities that have established and maintained Northrop Grumman's reputation and enabled technological differentiation in the marketplace. Dr. Romesser earned a PhD in physics from the University of Iowa. He is also a graduate of the USC Executive Management Program. He was elected a fellow of DEPS in 2002.

David J. Scott (Maj Gen, USAF, Ret.) retired from the Air Force in November 2012 after more than 34 years of service at the strategic, operational, and tactical level. His last assignment was as director, Operations, Strategic Deterrence, and Nuclear Integration, Headquarters U.S. Air Forces in Europe (USAFE), Ramstein Air Base, Germany. In this position he developed and implemented policy, obtained resources, and developed concepts of operation in order to ensure U.S. Air Forces in Europe are well organized, trained, and equipped. Those forces provide integrated air and missile defense, aeromedical evacuation, cyberspace operations, airfield operations and support, aerial refueling, combat aerial delivery, operational support airlift and Distinguished Visitor airlift, intelligence, surveillance and reconnaissance, and precision strike in the European and African Command area of responsibility, the only Major Command to have two Combatant Commands (COCOM) directly in their chain of command along with other Combatant Commands when tasked. Under his leadership, USAFE reorganized to be able to support the two COCOMs simultaneously. Gen Scott entered the Air Force in 1978, receiving his commission and degree through the U.S. Air Force Academy. He has served in a variety of positions at the squadron, group, wing, and joint levels in Europe, the Pacific, and the United States. Before his last assignment, he was director, Operational Capability Requirements, deputy chief of staff for Operations, Plans and Requirements, Headquarters U.S. Air Force, Washington, D.C. Gen Scott participated in the following schools: National Security Studies, Maxwell School of Government Syracuse, JFK School of Government, Harvard, and the Darden School of Leadership, University of Virginia, as well as the Combined Forces Air Component Course and Joint Flag Officer Warfight Course at Maxwell Air Force Base, Alabama. He is a command pilot with more than 3,000 flying hours in fighter aircraft, including approximately 200 combat hours.

Frank J. Serna is the director of systems engineering at Charles Stark Draper Laboratory. The Systems Engineering Directorate consists of approximately 200 engineers and 50 technicians and administrative staff, comprised of three divisions: Systems Engineering, Test and Evaluation, and Quality Assurance. The scope of projects includes the entire scope of Draper

Laboratory programs: Guidance Systems for Trident II, NASA Manned Space Programs, Missile Defense, Guided munitions, Maritime systems, low power electronics, and biomedical systems. He has more than 30 years of experience in organizations involved in contract research, development, and systems integration projects for national security sponsors. Mr. Serna has served on the Defense Science Board Task Force on Counter IED II. He is a steering committee member of the NDIA Systems Engineering division and the Massachusetts Advanced Cyber Security Center. Previously, Mr. Serna was the director of systems engineering in the Defense Enterprise Solutions Business Unit of Northrop Grumman and was director of software development in the Litton-TASC Business Unit. Finally, he was an original member of Missile Defense National Team for Systems Engineering and Integration. Mr. Serna holds a bachelor of science degree in engineering and applied science from Yale University and a master's degree in business administration from Northeastern University.

Andrew M. Sessler, a member of the National Academy of Sciences, is a Distinguished Emeritus Scientist and Distinguished Director, Emeritus, of the Lawrence Berkeley National Laboratory. He has published more than 300 scientific papers for which he has received many awards, including the Lawrence Award and the Wilson Prize. Dr. Sessler is a former president of the American Physical Society (APS), has been active in arms control and human rights, and has served as chairman of the Federation of American Scientists. He co-founded the human rights group SOS and was the first winner of the Nicholson Medal. Dr. Sessler served as chair of the 2000 Union of Concerned Scientists study *Countermeasures: A Technical Evaluation of the Operational Effectiveness of the Planned U.S. National Missile Defense System* and as a co-author of the 1987 APS report *The Science and Technology of Directed Energy Weapons*.

Appendix B

Terms of Reference

An ad hoc committee will plan and convene one workshop consisting of three 2-day meetings (spaced a month apart for logistical reasons) to: (1) consider directed energy technologies available to the Air Force; and (2) the potential application of directed energy technology to Air Force mission needs, with an emphasis on the defensive employment of directed energy capabilities against notional offensive weapons.

The committee will develop the agenda for the workshop, select and invite speakers and discussants and moderate the discussions.

In organizing the workshop, the committee might also consider additional topics close to and in line with those mentioned above. The meetings will use a mix of individual presentations, panels, breakout discussions, and question-and-answer sessions to develop an understanding of the relevant issues. Key stakeholders would be identified and invited to participate. One individually authored workshop summary document will be prepared by a designated rapporteur.

Appendix C

Workshop Session Agendas

SESSION 1
FEBRUARY 27-28, 2013
WASHINGTON, D.C.

Objectives

1. Introduce workshop committee members and participants
2. Receive briefings on current government directed energy technology initiatives
3. Plan for Session 2

FEBRUARY 27, 2013

CLASSIFIED SESSION

0900 **Welcome and Introductions**
Gen Ronald E. Keys (USAF, Ret.), Chair

0930 **Air Force Champion's Vision for Directed Energy**
Dr. Janet Fender, Chief Scientist, Air Combat Command (ACC)
Dr. Dave Robie, Office of the Chief Scientist, ACC

1030 Break

1045 **Air Force Research Laboratory (AFRL) Initiatives**
Dr. Robert Peterkin, Chief Scientist, Directed Energy Directorate, AFRL

1145 **Continue Discussions** *with Lunch Available*

1215 **High-Energy Laser Joint Technology Office (HEL-JTO) Initiatives**
Mr. Don Seeley, Acting Director, HEL-JTO

1315 **Army Initiatives**
Mr. Charlie Freund, Acting Chief, Directed Energy Branch, U.S. Army Research and Development Command, Armaments Research, Development and Engineering Center

END CLASSIFIED SESSION

UNCLASSIFIED SESSION

1415 **Army Initiatives**
Dr. Kip Kendrick, Chief, Directed Energy Division, U.S. Army Space and Missile Defense Command, Army Forces Strategic Command (VTC)

1515 Break

END UNCLASSIFIED SESSION

CLASSIFIED SESSION

1530 **Defense Advanced Research Projects Agency (DARPA) Initiatives**
Dr. Rich Bagnell, Strategic Technology Office, DARPA

1630 **Discussion of Day 1 Presentations**
All

1700 **Adjourn**

END CLASSIFIED SESSION

FEBRUARY 28, 2013

CLASSIFIED SESSION

0900 **Welcome and Introductions**
Gen Ronald E. Keys (USAF, Ret.), Chair

0910 **Office of the Secretary of Defense (OSD) Initiatives**
Dr. Spiro Lekoudis, Research Directorate, Defense Research and Engineering, OSD

1010 Break

1030 **Department of Energy Initiatives**
Dr. Vance Behr, Deputy Director for Military Systems and Technologies, Integrated Military Systems Center, Sandia National Laboratories

1130 **Continue Discussions** with Lunch Available

1230 **Missile Defense Agency (MDA) Initiatives**
Dr. Steve Post, Laser Technology Program Manager, MDA

1330 **Lincoln Laboratory Initiatives**
Dr. Dan Ripin, Assistant Leader, Laser Technology and Applications Group, MIT Lincoln Laboratory

1430 **Office of Studies and Analyses, Assessments and Lessons Learned, Headquarters U.S. Air Force (HAF/A9) Initiatives**
Mr. Jeffrey Saling, Chief, Mission Analyses Division, HAF/A9

1530 **Joint Non-Lethal Weapons Directorate (JNLDW) Initiatives**
Ms. Susan Levine, Principal Deputy Director (Policy and Strategy), JNLDW

1630 **Discussion of Day 2 Presentations/Plan for Session 2 (March 18-19)**
All

1700 **Adjourn**

END CLASSIFIED SESSION

SESSION 2
MARCH 18-19, 2013
WASHINGTON, D.C.

Objectives

1. Introduce workshop committee members and participants
2. Receive briefings on current government and industry directed energy technology initiatives
3. Plan for Session 3

MARCH 18, 2013

CLASSIFIED SESSION

0900 **Welcome and Introductions**
Gen Ronald E. Keys (USAF, Ret.), Chair
Dr. Janet Fender, Chief Scientist, ACC

0915 **Takeaways from Meeting 1**
Dr. Robert Katt, Rapporteur

1015 Break

1015 **Barriers to Transitioning Technologies for Operational Employment**
Mr. Terry Jaggers, Air Force Studies Board Director, Former Deputy Assistant Secretary of the Air Force for Science, Technology, and Engineering

1115 **Results of Directed Energy Task Force**
Maj Gen David Scott (USAF, Ret.), Workshop Committee Member

1215 **Continue Discussions** *with Lunch Available*

1300 **Navy Initiatives**
Mr. Peter Morrison, Research Program Officer, Office of Naval Research

1400 Break

1415 **U.S. Pacific Command (PACOM)**
Mr. Kenneth Bruner, S&T Advisor, PACOM

1515 **Discussion of Day 1 Presentations**
All

1700 **Adjourn**

END CLASSIFIED SESSION

MARCH 19, 2013

CLASSIFIED SESSION

0900 **Welcome and Recap of Day 1**
Gen Ronald E. Keys (USAF, Ret.), Chair

0930 **Boeing Initiatives**
Mr. Keith Coleman, CHAMP Program Manager, Phantom Works

1030 Break

1045 **Raytheon Initiatives**
Dr. Peter Duselis, Manager, High Power Microwave Systems Department, Pulsed Power Division, Raytheon Ktech

1145 **General Atomics Initiatives**
Mr. Chris Pehrson, Strategic Development Director, Laser and Electro-Optics, General Atomics

1245 **Continue Discussions** with Lunch Available

1315 **Navy Initiatives**
Mr. Tom Glover, Head, Emergent Weapon Systems, GS-15/NM5 and Lead, Directed Energy Weapons, NAVAIR

1415 **Technological Assessment of Foreign Directed Energy**
Mr. Mike Bernard, Senior Intelligence Analyst, AFRL

1515 Break

1530 **Discussion of Day 2 Presentations/Plan for Session 3 (April 24-25)**
All

1700 **Adjourn**

END CLASSIFIED SESSION

SESSION 3
APRIL 24-25, 2013
WASHINGTON, D.C.

Objectives

1. Introduce workshop committee members and participants
2. Receive briefings on current government and industry directed energy technology initiatives
3. Review draft workshop summary
4. Disucss strategy for outbriefing trail

APRIL 24, 2013

CLASSIFIED SESSION

0830 **Welcome and Introductions**
Gen Ronald E. Keys (USAF, Ret.), Chair

0835 **Takeaways from Meeting 2**
Dr. Robert Katt, Rapporteur

0930 **Air-Sea Battle (ASB) Concept of Operations (CONOPS)**
Col Jordan Thomas, HAF/A5X
Mr. William Dries, HAF/A5X (Air-Sea Battle Office)

1. *Which service has the lead for protecting such fixed bases as Guam?*
2. *Has this been coordinated with the Army for their base protection?*
3. *Are directed energy weapons (DEW) being considered in the trade space for that protection?*
4. *What ASB operational needs could potentially be met by DEW?*
5. *Have any threats been identified by ASB that suggest potential BlueForce DEW applications?*

1030 Break

UNCLASSIFIED SESSION

1045 **Results of Army DEW Analysis of Alternatives (AoA) (VTC)**
Dr. Kip Kendrick, Chief, Directed Energy Division, U.S. Army Space and MissileDefense Command, Army Forces Strategic Command

1. *Describe the AoA that was done for the counter rocket, artillery, and mortar base defense mission and the outcome of the AoA.*
2. *Why was the final decision for the go-ahead not the one that came out on top for the AoA?*
3. *Summarize what the AoA found were the advantages of DEW overalternative kinetic solutions.*

1145 **Continue Discussions** *with Lunch Available*

END UNCLASSIFIED SESSION

CLASSIFIED SESSION

1215 **DARPA Initiatives**
Dr. Joe Mangano, Program Manager, Advanced Lithography Program, Microsystems Technology Office

1. *Describe your laser program including its progress to date and its future plans as well as its relevance to the Air Force.*
2. *Provide examples of Air Force operational needs that could possibly bemet by low- and/or medium-power lasers.*

1315 **Lockheed Martin Laser Efforts**
Mr. Tom Burris, Lead Systems Engineer, Directed Energy Program, Lockheed Martin

1. *Describe fairly near-term (5 years or less) laser DEW concepts that address Air Force missions. Include effectiveness for Air Force missions and CONOPS thoughts. To this end, the workshop participants do not need details about the technology, per se. These missions could involve low-to-medium-power lasers (e.g., 10 or 10's of kW power levels). NOTE: HEL-JTO, MDA, AFRL, MIT/LL, DARPA HELLADS, Army SMDC, Navy ONR, ACC and General Atomics have already provided briefings. The workshop participants would prefer that the briefing not be proprietary since a summary of the briefing will go into a National Academy document. The presentation can be up to the Secret level.*

1415 **Northrop Grumman Laser Efforts**
Mr. Kenton Ho, Director of Programs for Laser Products, Northrop Grumman

Guidance same as for Lockheed Martin

1515 Break

1530 **Boeing Airborne DEW Systems**
Mr. Michael Black, Business Development Manager, Boeing

Guidance same as for Lockheed Martin

1630 **General Discussion on Day 1 Presentations**
All

1700 **Adjourn**

END CLASSIFIED SESSION

APRIL 25, 2013

CLASSIFIED SESSION

0900 **Welcome and Introductions/Recap of Day 1**
Gen Ronald E. Keys (USAF, Ret.), Chair

0915 **ACC CHAMP Briefing**
Dr. David Robie, Office of the Chief Scientist, ACC

1015 Break

1030 **Continue Discussions**
All

1200 **Continue Discussions** with Lunch Available

1300 **Continue Discussions**
All

1700 **Adjourn**

END CLASSIFIED SESSION

APPENDIX D

WORKSHOP PARTICIPANTS

SESSION 1
FEBRUARY 27-28, 2013
WASHINGTON, D.C.

Workshop Planning Committee

Gen Ronald E. Keys, (USAF, Ret.), *Chair*
Dr. William L. Baker
Dr. Jacqueline G. Gish (NAE)
Mr. Matt L. Mleziva
Dr. Thomas E. Romesser (NAE)
Maj Gen David J. Scott (USAF, Ret.)
Mr. Frank J. Serna
Dr. Andrew M. Sessler (NAS)

National Research Council Staff

Mr. Terry Jaggers, AFSB Director
Mr. Carter Ford, Program Officer
Dr. Robert Katt, Rapporteur
Ms. Marguerite Schneider, Administrative Coordinator

Speakers

Dr. Janet Fender, Chief Scientist, ACC
Dr. David Robie, Office of the Chief Scientist, ACC
Dr. Rich Bagnell, Strategic Technology Office, DARPA
Dr. Vance Behr, Deputy Director for Military Systems and Technologies, Integrated Military Systems Center, Sandia National Laboratories
Mr. Charlie Freund, Acting Chief, Directed Energy Branch, U.S. Army RDECOM-ARDEC

Dr. Kip Kendrick, Chief, Directed Energy Division, U.S. Army Space and Missile Defense Command, Army Forces Strategic Command
Dr. Spiro Lekoudis, Research Directorate, Defense Research and Engineering, OSD
Ms. Susan LeVine, Principal Deputy Director (Policy and Strategy), JNLWD
Dr. Robert Peterkin, Chief Scientist, Directed Energy Directorate, AFRL
Dr. Steve Post, Laser Technology Program Manager, MDA
Dr. Dan Ripin, Assistant Leader, Laser Technology and Applications Group, MIT Lincoln Laboratory
Mr. Jeffrey Saling, Chief, Mission Analyses Division, HAF/A9
Mr. Don Seeley, Acting Director, HEL Joint Technology Office

Guests

Dr. Lawrence Grimes, Directed Energy Staff Specialist, Research Directorate, Defense Research and Engineering, OSD
Mr. John Haynes, Deputy for Directed Energy, Office of the Deputy Assistant Secretary of the Air Force for Science, Technology, and Engineering

SESSION 2
MARCH 18-19, 2013
WASHINGTON, D.C.

Workshop Planning Committee

Gen Ronald E. Keys, (USAF, Ret.), *Chair*
Dr. William L. Baker
Dr. Jacqueline G. Gish (NAE)
Mr. Matt L. Mleziva
Dr. Thomas E. Romesser (NAE)
Maj Gen David J. Scott (USAF, Ret)
Mr. Frank J. Serna
Dr. Andrew M. Sessler (NAS)

National Research Council Staff

Mr. Terry Jaggers, AFSB Director
Mr. Carter Ford, Program Officer
Dr. Robert Katt, Rapporteur
Ms. Marguerite Schneider, Administrative Coordinator

Speakers

Mr. Mike Bernard, Senior Intelligence Analyst, AFRL

Mr. Kenneth Bruner, S&T Advisor, PACOM
Dr. Keith Coleman, CHAMP Program Manager, Boeing Phantom Works
Dr. Peter Duselis, Manager, High Power Microwave Systems Department, Pulsed Power Division, Raytheon Ktech
Dr. Tom Glover, Head, Emergent Weapon Systems, GS-15/NM5, and Lead, Directed Energy Weapons, NAVAIR
Mr. Peter Morrison, Research Program Officer, ONR
Mr. Chris Pehrson, Strategic Development Director, Laser and Electro-Optics, General Atomics

Guests

Dr. David Robie, Office of the Chief Scientist, ACC
Mr. John Haynes, Deputy for Directed Energy, Office of the Deputy Assistant Secretary of the Air Force for Science, Technology, and Engineering

SESSION 3
APRIL 24-25, 2013
WASHINGTON, D.C.

Workshop Planning Committee

Gen Ronald E. Keys, (USAF, Ret.), *Chair*
Dr. William L. Baker
Dr. Jacqueline G. Gish (NAE)
Mr. Matt L. Mleziva
Dr. Thomas F. Romesser (NAE)
Maj Gen David J. Scott (USAF, Ret.)
Mr. Frank J. Serna
Dr. Andrew M. Sessler (NAS)

National Research Council Staff

Mr. Terry Jaggers, AFSB Director
Mr. Carter Ford, Program Officer
Dr. Robert Katt, Rapporteur
Ms. Marguerite Schneider, Administrative Coordinator

Speakers

Dr. David Robie, Office of the Chief Scientist, ACC
Mr. Tom Burris, Lead Systems Engineer, Directed Energy Program, Lockheed Martin
Mr. William Dries, Air-Sea Battle Office
Mr. Kenton Ho, Director, Laser Products Directed Energy, Northrop Grumman

Dr. Kip Kendrick, Chief, Directed Energy Division, U.S. Army Space and Missile Defense Command, Army Forces Strategic Command
Dr. Joe Mangano, Program Manager, MTO, DARPA
Dr. Harold Schall, The Boeing Company
Col Jordan Thomas, Air-Sea Battle Office

Guests

Mr. John Haynes, Deputy for Directed Energy, Office of the Deputy Assistant Secretary of the Air Force for Science, Technology, and Engineering
Mr. Cornelius Morgan, Booz Allen Hamilton
Mr. Douglas Raaberg, Northrop Grumman